Voices for Climate Independent Food

"Without action at the global level to address climate change, we will see farmers across Africa – and in many other parts of the world, including in America – forced to leave their land. The result will be mass migration, growing food shortages, loss of social cohesion and even political instability."   Kofi Annan, Former UN Secretary- General

"There are reasons to expect more frequent food price spikes, given that it will be more common to see weather conditions that are considered extreme." David Lobell, Assistant Professor of Environmental Earth System Science, Stanford University.

"When food prices rise sharply, families cope by pulling their kids out of school and eating cheaper, less nutritious food, which can have catastrophic life-long effects on the social, physical and mental wellbeing of millions."   Jim Yong Kim, the World Bank Group President.

"Of course I feel hungry. I feel hungry until I become weak. When I'm hungry, if possible, I prepare a broth for myself and my kids – otherwise we drink some water and we sleep." Adjitti Mahamat, from Chad, where 3.6 million people are food-insecure due to drought, chronic poverty, and food prices, which have increased 60% in the last two years.

Over 46 million Americans receive food stamps because they are poor and hungry. Over 50 million people in the U.S. are food insecure and over 70 million receive food support.   USDA, September 2012.

"This year's heat and drought turned our family's dreams to dust."
Canadian farmer.

"We are dealing with a food system here. There is a whole chain that is also going to be affected by climate change." Professor John Porter, University of Copenhagen.

Extreme weather events in a single year could bring about price spikes of comparable magnitude to two decades of projected long-run price increases. Oxfam Issue Briefing, *Extreme Weather, Extreme Prices, The costs of feeding a warming world*, September 2012.

Climate Independent Foods

Survive and Thrive with Freedom from Weather, Water and Waste

Mark R. Edwards

Robert Henrikson

Sustainable food production that saves
fossil resources and cleans polluted ecosystems.

The Green Algae Strategy Series

AlgaeAlliance.com

AlgaeCompetition.com

Key words.

Food	Microfarm	Sustainability	Abundance
Water	Nutrient recovery	Environment	Hunger
Regenerative	Organic farming	Smartcultures	Poverty
Agriculture	Climate change	Energy	Drought
Aquifers	Micronutrients	Freedom food	Ecosystems
Nutralence	Global awareness	Algaculture	Malnutrition
Genetics	Renewable energy	Biotechnology	Pollution
Microalgae	Industrial farming	Ecology	Algae

ISBN-13: 978-1479276844
ISBN-10: 1479276847

BISAC: Science / Biotechnology

Copyright © 2012 by Mark R. Edwards and Robert Henrikson

Cover photo copyright © 2012 by Mark R. Edwards.

Climate Independent Foods may be used for educational purposes.

Dedication

To our children who may enjoy a legacy of abundance with food security and clean, green ecosystems.

Contents

Chapter 1. Weather, Water and Waste *1*

Chapter 2. What will our Children do for Food? *19*

Chapter 3. What is the Food Bill of Rights? *27*

Chapter 4. What is Abundance? *37*

Chapter 5. What are Freedom Foods? *46*

Chapter 6. Algae in Human Food History *66*

Chapter 7. What makes Algae Special? *75*

Chapter 8. What are Peace Microfarms? *83*

Chapter 9. Microfarm Examples *101*

Chapter 10. Why not Freedom Foods Now? *117*

Chapter 11. e-Footprint and Ecobalance Diet *129*

Chapter 12. Future Foods for any Climate *137*

Acknowledgements *153*

Appendix I Climate Change Impacts on Food *154*

The Green Algae Strategy Series *163*

Climate Independent Foods

Forward

Climate chaos puts our food supply and human societies in jeopardy. Each recent year temperatures rise. New records were set in 2012 for searing heat and drought, the number and severity of wildfires, extreme storms, floods, West Nile deaths, sea ice melt and crop insurance loss. The USDA declared over half of all counties in the U.S. natural disaster areas. Heat and drought engulfed 90% of corn and soy acreage. Climate experts suggest the extreme weather creates a new norm that will only become more severe for field crops.

Climate independence creates a strategic food debate. Should we bet our children's lives on industrial foods that fail with weather events or scarce water? Should we continue growing foods that consume massive amounts of fossil resources and then create ecological suicide with waste and pollution?

Alternatively, can we design a food supply that assures growers can produce 20 to 30 times more food per acre every year, independent of weather? Can we build a food supply sustainable for at least seven generations with abundance methods that provide superior health and taste with freedom from weather, water and waste?

Human caused pollutants have intensified the greenhouse effect and destabilized the climate so rainfall, temperatures and storm severity are no longer predictable from past trends. Failing a novel solution, climate chaos will decimate crops, drive food costs higher and further amplify local, regional and world hunger.

Weather, water and waste converge to form the Achilles Heel of modern agriculture. The arrow that fatally wounded Achilles was a lucky shot. The arrow that threatens human societies – climate chaos – already flies in the air. Its validity has consensus support from 98% of meteorological scientists. You may feel its weather disruptions when you walk out your door.

Multiply your discomfort about extreme weather events by 10 and you may approach the impact on food crops. Plants have far less weather tolerance than people. They are bound to the soil and cannot move. Plants are dependent on the reach of their roots for water and nutrients. Crops produce food within a very narrow range of weather, water, nutrients and soil fertility. Disruption to only one of the multiple variables vital to crops first diminishes and then destroys the fruit of the vine, the food.

Food security discussions typically focus on productivity, tolerance and resilience. Food productivity increases peaked in the 1980s and have been decreasing ever since. A few large companies like Monsanto hope to bring genetically engineered, (GE) drought and heat tolerant seeds to the market. Who knows whether GE seeds will grow reliably with climate chaos? Others advocate resilient farming with small organic farms. Tolerance and resilience are fine objectives that hold promise for incremental crop improvements but do not offer freedom from weather, water or waste.

Human survival depends on new sustainable solutions that free farmers from weather constraints and allow food production in nearly any climate. Increasing demand and water scarcity will ignite more water wars. Farmers will lose water wars. They need freedom from dependence on enormous amounts of fresh water. Growers need freedom from increasingly costly fossil resources and the waste and pollution those resources impose on their fields and the environment.

Abundance methods reinvent our food supply by growing food and other forms of energy with plentiful resources that will not run out. Abundance does not compete with industrial foods because growers use renewable rather than fossil resources. Abundance and industrial food production are compatible and can produce simultaneously. Expanding populations need every available form of food production.

Abundance enables growers to produce climate independent foods by cultivating the most productive crop on our planet – algae. Growing food organically, low on the food chain, yields sustainable advantages for consumers, growers and ecosystems. Abundance methods offer the opportunity to leave a superb legacy for our children – green fields, clean streams and affordable healthy food.

Climate Independent Foods

Key Terms

Fossil foods are unsustainable commercial foods that escalate with fuel prices. Industrial foods suffer from hidden hunger, deliver empty calories, fail in bad weather, are dependent on costly non-renewable resources and create massive waste and pollution.

Climate independent foods produce 20 to 30 times more food per acre every year, free of weather, water or waste. Growers use abundance methods to grow fossil-free foods that assure a sustainable food supply for many generations.

Abundance methods offer an alternative for growing food and other forms of energy with plentiful resources that are cheap and will not run out – primarily solar energy, CO_2 and sterilized waste stream nutrients. Abundance allows sustainable and affordable organic food production in any practically any altitude, latitude or geography.

Freedom foods are climate independent foods that redesign our food supply from the foundation of the food chain. Freedom foods free consumers for smart food choices, growers from fossil resource cost and consumption and ecosystems of waste and pollution. Freedom foods include the full spectrum of microorganisms such as algae, yeast, fungi, bacteria, archaea, protists, plankton and many others.

Nutralence are an attribute of freedom foods that have 200% more nutrient availability and 200 to 500% more nutrient density per bite than industrial foods. Foods with natural nutralence do not deliver the empty calories that cause obesity and diabetes.

Peace microfarms avoid conflicts over diminishing freshwater and other fossil resources. Microfarms can be sited practically anywhere. Growers recover low cost nutrients from sterilized waste streams and transform them into valuable freedom foods and other products.

Microcrops are foods low on the food chain that include algae and the microorganisms algae attract such as fungi, bacteria, viruses and yeast. Microcrops offer many sustainable advantages over field crops.

Chapter 1. Weather, Water and Waste

The impact of climate change on food production can already be seen, and will worsen as climate change gathers pace. —David Lobell, Stanford University.

Food enables life. Calories rich in proteins and nutrients provide the vital of energy for plants, animals and humans. Our vigor comes from the solar energy transformed to chemical energy in green plant bonds stored in plants, or the animals that eat plants. Our bodies burn calories stored in food to generate our vitality.

Three factors put our food supply in jeopardy, weather, water and waste. Human actions have magnified the threat to food security. Greenhouse gasses create climate chaos, devastating crops. Overconsumption of water leaves wells dry and fields scorched. Agricultural waste pollutes fields and ecosystems and creates serious health problems for people and animals that live downstream or downwind. Farm erosion and waste systemically degrade croplands and leave huge dead zones in lakes, estuaries and oceans.

What if we could grow healthier foods:
- Independent of weather or climate?
- Without fresh water or other fossil resources?
- Sustainably, recycling nutrients without waste?
- Avoid erosion and pollution and instead clean air and water?
- Remediate degraded cropland?

Climate Independent Foods

Abundance offers a novel path for healthier people, producers and our planet.[1] Current natural and human events make abundance strategically critical.

Food chain

The food chain supplies energy for every living creature. Humans consume, directly or indirectly, about 40% of all the solar energy flowing through the food chain, which makes the food network critical for human societies.[2] The food chain is robust in its breadth and diversity but has two serious weaknesses for sustaining human life:

1. Should any one of the links in the food chain fail, a single trophic level, the entire food system could collapse. Higher tropic level creatures such as animals and humans are most at risk since they are at the top of the food chain and are dependent on numerous lower links in the chain.

2. The food chain is inefficient and transfers only about 10% of the energy from a lower level to the next higher. Therefore, huge amounts of biomass must be produced to feed animals and even more to feed humans. Trophic chains are so inefficient in energy transfer that fish make up less than 1% of human nutrition.[3]

Human overconsumption – removing too many resources from the food web – amplifies food chain collapse risk. Climate chaos adds a serious threat to global food supplies. Water scarcity devastates large food supply sectors. The additive impacts of waste, erosion and pollution are expanding dead zones in rivers, lakes, estuaries and oceans where no organisms can survive. The 30% of fertile land abandoned over the last 40 years due to the extractive nature of industrial agriculture are effectively dead zones.

When local food supplies ran low for hunter-gatherers, nomadic tribes simply moved to food. Modern farmers grow food far distant from cities and invest roughly half their fossil energy in processing, packaging and transporting food to concentrations of consumers.

Hunger today

The 2010 FAO State of the World estimates 925 million hungry people in the world. This represents 13%, or almost 1 in 7 people are hungry. The larger problem is that half the people on Earth are food insecure, without reliable access to good food.

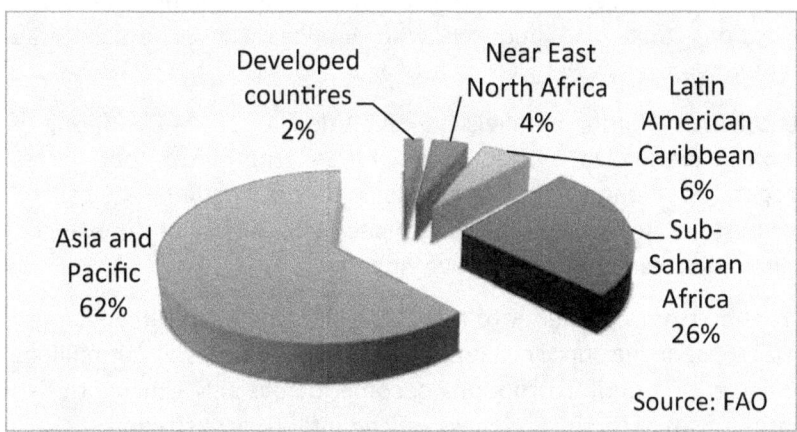

Figure 1.1 Hungry People in 2010 = 925 million

In the U.S., where we are blessed with ideal growing conditions, one out of five Americans – over 60 million citizens – receives food support because they are food insecure and hungry.[4] Food support, food stamps and school lunch programs, provides about $1 per meal. Dollar meals eliminate the opportunity for families and their children to participate in the American dream. Poor Americans lack food freedom. They must feed their children cheap, nutrient deficient foods that cause obesity, fatigue, dull brains and diabetes.

One out of every four American children struggles to get enough food for their bodies and minds to develop properly.[5] Food insecurity forces our children to eat cheap food, which typically delivers loads of fat, salt and cholesterol, but poor taste and color. These foods deliver empty calories, largely devoid of nutrition. Malnourished children incur developmental impairments that limit their physical, intellectual and emotional development.[6] Developmentally impaired children place a huge drag on families, medical facilities and education.

Climate Independent Foods

The structure of our food supply should be a top priority for national defense, security, health and education. Politicians and policy leaders ignore the strategic value of our food supply – to our peril.

The French Revolution in 1784 ignited due to escalating food prices. Crops failed due to extreme weather from El Niño, amplified by the 1783 volcanic activity at Laki and Grímsvötn, Iceland. In *Empires of Food*, Evan Fraser and Andrew Rimas document numerous civilizations that rose and fell with famines caused by extreme weather events.

The 2011 Arab Spring parallels the revolutions of 1848. The Spring of Nations experienced a year of revolutions that broke out across Europe. The "hungry '40s" saw a decade bad weather that caused poor harvests across Europe. Hungry people in Europe became angry. Angry people can bring down governments.

Disruption from weather or to any of the many resources necessary to grow crops, or the supply chains that distribute food, will put millions of people in in peril. Disruptions become increasingly likely with bad weather, water scarcity and massively pollutive waste streams.

Weather, water and waste

Modern industrial food production is non-sustainable due to three fatal errors, weather, water and waste. Food crops:

- **Fail with bad weather.** Many weather anomalies can degrade or destroy a crop during any of the 120 days of its growing cycle. Climate chaos will intermittently devastate food crops in many growing regions. The economics of our food supply cannot afford the risk of crops that fail too often.

- **Fail with insufficient fresh water.** Food crops consume vast amounts of fresh water, which must be delivered on time. Crops in temperate regions consume about a million gallons and acre and in hotter climates such as the western U.S., two to three million gallons an acre. Farmers are pumping water from aquifers at six to 100 times replacement rates.

- **Fail from waste streams**. Cultivation and applications on modern farms are so disruptive that it damages itself. Industrial farming extracts soil nutrients and humus, continually degrading soils. Salt invasion from irrigation destroys fertility and prevents germination. Constant erosion from wind and water degrade cropland until it has to be abandoned. Eroded nutrients and agricultural poisons pollute fields, air and surface and well water.

Modern agriculture appears to be in a race to find which threat crashes the food supply first. Currently, climate chaos receives the most notice because hotter temperatures and more severe storms make visual impact in TV news. Water scarcity and quality are probably more severe threats to the food supply, especially when aquifers crash in the near future. Agricultural erosion and pollution jeopardizes food production with the insidious loss of soil and fertility and the build up of pollutants such as salts and poisons.

In spite of political debates, the U.S. continues to increase rather than curb greenhouse gasses the cause global warming. The Inventory of U.S. greenhouse gas emissions tracks the national trend in greenhouse gas emissions. The report shows that in 2010, U.S. greenhouse gas emissions totaled 6,900 million metric tons CO_2 equivalent. Since 1990, U.S. greenhouse gas emissions have increased by 10.5%. Neither legislation nor apparent political will exists to moderate air, water or soil pollution. Some good laws have passed like the clean air and clean water acts but they lack enforcement.

Agriculture

Agriculture consumes over 12 billion acres of land globally with about 70% in pasture and 30% in crops.[7] About 11,000 years ago, humans began supplementing available foods with agriculture. The husbandry of food plants enabled humans to diversify their diets and spend less time insuring their food supply. For 99% of farming history, farmers cultivated their fields with organic production. They grew a diversity of plants, rotated crops to replace organic nutrients, avoid erosion and minimize pests, applied animal and green manure for fertilization to replenish the soil nutrients. Animal manure comes from draft and other farm animals that recycle organic carbon while green manure comes from a cover crop that is ploughed under to replace soil nutrients, add rich organic matter and support water retention.

Climate Independent Foods

For thousands of years agriculture was environmentally benign because farmers relied on natural ecological processes. Crop residues were incorporated into the soil or fed to livestock. Manure was returned to fields where it replaced nutrients and soil organics. Typical mixed production farms with crops and animals were closed, stable and sustainable ecological systems that generated few external impacts. Mixed production also conserved vital nutrients.

Modern industrial agriculture began in 1950 after World War II as a plan to provide cheap food to the world. Early productivity increases of 3.5% were superb but have slowed to less than 1.5% over the last two decades and are decreasing. Today, four times more people are hungry globally than in 1950. Even more threatening, all the technology embedded in modern farming makes the food supply more vulnerable to severe weather events.

Environmental catastrophes

Industrial agriculture improved human societies but weather, water and waste streams put modern food supplies at severe risk. When weather, water or fertility failed, communities and occasionally entire civilizations, perished. Before the last people died from community starvation, war and illness decimated the population.

Community starvation must be the most agonizing way for humans to die. Families are forced to watch helplessly while their weakest suffer. They see their children and elders in prolonged and excruciating misery while the human body consumes its own tissues in a desperate attempt to sustain energy. The victim's skin changes color and loses elasticity while the stomach loses its ability to digest. Eyes sink in their sockets and victims lose their memory and become weak, fatigued and disoriented. Death usually comes from diseases such as dysentery, pneumonia or heart failure that mercilessly attack the weakened body.

Elizabeth Kolbert in *Field Notes from a Catastrophe* chronicled a list of sophisticated cultures that sustained themselves for hundreds of years and then crashed due to overconsumption of critical resources and climate change, primarily multiyear droughts, such as:

1. Tiwanaku, Lake Titicaca in the Andes – crash: A.D. 1100, drought
2. Classic Mayan civilization – crash: A.D. 800, drought
3. Old Kingdom of Egypt – crash: 2,200 B.C., drought
4. Akkadian empire – crash: 2,200 B.C., drought[8]

Drought may be a climate event but droughts amplify the need for irrigation. Irrigations systems designed for normal years are unlikely to deliver sufficient water in hot years when crops may need two or three times more water. Heat also intensifies salt as evaporation of irrigation water leaves a white soil crust of irrigation salts that diminish and eventually destroy soil fertility.

Food crops are voracious consumers. A ton of grain consumes 1,000 tons of water of water. The U.S. is the largest exporter of wheat, corn and soy to the world. When the U.S. exports a ton of wheat, buyers are effectively buying 1,000 tons of water. What are the ethics of hiding the cost of 1,000 tons of water in each ton of exported grain when many of our children will face water scarcity?

Farmers must cultivate crops for the entire growing season before harvest. If weather, water or fertility problems occur late in the growing season, the farmer loses not just the crop but also all the labor, money and natural resources invested. The farmer also misses the opportunity to grow a short-season crop to feed the family.

Jared Diamond in *Collapse: How Societies Choose to Fail or Succeed* describes similar eco-meltdowns that caused the Anasazi of the U.S. Southwest and the Viking colonies of Greenland to crash.[9] He shows how patterns of population growth combined with drought, over-farming and destruction of natural resources leads to deforestation, erosion and starvation.

Climate chaos

Climate chaos causes food insecurity. People will do almost anything to secure food for themselves and their family. Food stress ignites food riots that may escalate to war Over 40 nations endured food riots in 2008 due to food scarcity and affordability. Food bandits stole food stores and stripped farmers' fields. Some countries had to use their armed forces to distribute food.

Climate Independent Foods

The 12 months June 2011 to June 2012 were the hottest since modern recordkeeping began in 1895 and July the hottest month on record. Additive to heat, the worst drought in five decades cut grain production 35%. Drought and heat more than doubled corn prices to record highs above $8 a bushel. The USDA extended drought aid to half of all counties, a total of 1,297 spanning 29 states that have been designated natural disaster areas. Hot, dry weather also hit crops in southern Europe.

The Intergovernmental Panel on Climate Change SREX 2012 report indicates that extreme weather events are becoming more common.[10] Several studies have shown that over 98% of climate scientists actively doing research agree that human-caused climate change is happening.[11]

> *September 2012 was the 331th consecutive month with a global temperature above the 20th century average.*

Heat waves, coastal flooding, extreme precipitation and droughts are increasing. Higher ocean temperatures transfer more energy and warmer air carries more water vapor to weather systems, resulting in fiercer storms. Downpours are increasing which causes massive runoff, erosion and pollution.

Extreme weather events are the new normal

Agricultural Secretary Vilsack predicted higher grain prices would escalate meat and dairy prices for years. President Barack Obama recognized the worst drought in 50 years and directed the USDA to buy $170 million in meat to help farmers.

Weather, Water and Waste

Temperature spikes and drought, combined with insufficient water, caused crop losses in North Africa in 2011. Several revolutions in North Africa ignited because the dictators were unable to supply their citizens with sufficient food. People in Tunisia and Egypt were tolerant of their tyrants until families could not find affordable food. Food riots exploded into revolution, which toppled governments and ousted the dictators.

Too much heat devastates food grain yields. As temperature rises, the rate of photosynthesis increases to about 68° F and then plateaus up to 95 ° F. Photosynthetic activity declines above 95° F and stops at 104° F.[12] Rice, wheat and corn cannot pollinate above 104° F, which leads to crop failure. Combinations of heat, dry winds and insufficient soil moisture create partial pollination problems below 104° F.

Heat stress causes plants to curl their leaves to get less solar exposure and the stomata to close on the underside of the leaves in order to decrease moisture loss. Both actions disrupt photosynthesis and send the plant into thermal shock. David Lobel and Gregory Asner analyzed 16 years of corn and soybean data across 618 countries and concluded that each 1.8° F rise in temperature, over normal, caused a yield decline of 17%.[13] Heat stress causes yield drag for most food crops of 5—25% but severe heat stress from only a few extra degrees may destroy an entire crop. Heat stress breaks crop vitality and leaves the plants more vulnerable to disease, pests and fungus.

The heat wave in Europe in 2003 was 3.6° C above the long-term climatology norm and led to the deaths of an estimated 52,000 people.[14] Italy experienced a record drop in maize yields of 36% from a year earlier. In France maize and fodder production fell by 30%, fruit harvests declined by 25%, and wheat harvests declined by 21%.[15]

Temperature spikes, both high and low, can be extremely destructive especially at critical phases such as germination, early growth, pollination and fruiting. High winds associated with fierce storms can destroy a crop or knock stocks around so they are twisted and harvesters cannot cut through them.

Higher mean temperatures are only part of the story as higher low temperatures can be damaging. Some food crops such as pit fruits –

peaches, plums and apricots – require a certain number of cold hours, typically 450 hours below 44°F. If the weather stays warm over winter and receives too few cold hours, the tree buds out in the spring but the buds fall off and the tree fails to set fruit. A fruit orchard takes five years to grow trees that produce fruit so tree farmers cannot just change to a different crop the following year.

Secretary of Energy Steven Chu predicts California's farms and vineyards could vanish by the end of the century and its major cities could be in jeopardy if Americans do not act to slow the advance of global warming.[16] He also predicted that 90% of the Sierra Nevada snow pack on which California cities and agriculture depends would be gone by the end of the century.

Climate scenarios for 2020 project that Mexico will lose over one million acres of maize production to hotter temperatures.[17] Corn production in the U.S. will also be forced to shift north. Temperatures in Vermont may be too warm to produce maple syrup, depriving farmers of their livelihoods.

David Battisti and Rosamond Naylor used data from 23 global climate models contributing to the Intergovernmental Panel on Climate Change's 2007 scientific synthesis to show with a greater than 90% probability that growing season temperatures in the tropics and subtropics by the end of the 21st century will exceed the most extreme seasonal temperatures recorded from 1900 to 2006. In temperate regions, the hottest seasons on record will represent the future norm in many locations. More than 3 billion people live in the tropics and subtropics, many of whom live on under $2 per day and depend primarily on agriculture for their livelihoods.[18] Their analysis portends catastrophic loss of food supplies.[19]

More heat will compound food insecurity caused by variable rainfall. Heat will increase the incidence of agricultural droughts caused by elevated evaporation from soils, transpiration from plants, low soil moisture and high rates of water runoff from hardpan soils. Excess heat causes virga – rain that evaporates before it hits the ground.

The additional heat prompted the USDA to publish a new plant hardiness zone map for gardeners. The map was outdated by the time it was published and the USDA withdrew the map. Professor Nir Krakauer from City College New York published an on-line hardiness map that includes recent data showing winter temperatures are increasing more rapidly than summer temperatures. The map includes an online calculator for longitude and latitude to see the adjusted temperature change.[20] The map shows gardeners can grow varieties 100 miles further north than only 30 years ago.

Extreme weather events can destroy bridges and roads, making food transport problematic. Heat multiples human pests such as mosquitos and scorpions. The CDC reported that 2012 set the record for West Nile Virus outbreak, range and deaths, which exceeded 1200.

Hotter temperatures and especially hotter winters have extended the fire season nearly three months. Compared with 40 years ago, large wildfires are seven times more likely. The total area already burned in 2012 is 30% more than in an average year. Fires have consumed more than 9 million acres, an area larger than the state of Maryland.

U.S. Fires in 2012 **Large U.S. Fires 1970 - 2011**

 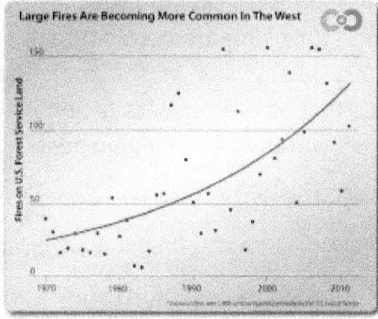

Source: U.S. Forest Service 2012

Pests proliferate in heat including thrips, weevils, white flies and invasive insect species. Mosquitos are vicious predators, causing over 1,600 West Nile cases and 66 deaths in 2012, which set new records. Heat stresses crops and saps their resistance to pests and crop diseases. Other pests infest stored foodstuffs while yeast, bacteria, and fungi can spoil food quickly. Appendix I provides a summary of climate chaos impacts on food production.

Climate Independent Foods

Water scarcity

Worldwide, 70% of all the water diverted from rivers or pumped from underground is used for irrigation.[21] Industry uses 20% and residential users consume 10%. In the western U.S., agriculture draws nearly 80% of the water. With the demand for water growing steadily in all three sectors, competition is intensifying. In this contest for water, farmers almost always lose to money – cities and industry.[22]

Roughly 60% of the world's croplands depend on extracted water for irrigation. Over 70% of farmers extract water six to sixty times the natural replacement rates. Farmers individually and collectively act out the tragedy of the commons.[23] Each farmer acts independently and rationally maximizing self-interest and overdrawing water from the public well or aquifer. Everyone knows the aquifer will crash but each farmer acts to get as much water as they need now. They fear that if they do not, others will take "their" water. Many aquifers will run dry within 20 years as too many straws draw too much water.

Water will become increasing scarce with global warming because melting snow packs and glaciers will cause rivers to run dry. More severe storms mean less water captured during runoff. Hot dry winds cause a faster loss of soil moisture, increasing the need for more water. Just as are farmers facing water scarcity, cities are running out of fresh water and driving up costs.

All the water that will ever be is, right now.

- National Geographic

Over half the crops by value in the U.S. are grown on land with insufficient rain that requires irrigation. Nearly all croplands from 100 miles west of the Mississippi River (the 100th Meridian) to the Pacific Ocean need irrigation. America's croplands are extremely productive with sufficient water because they receive considerable solar energy that enhances photosynthesis. While extra sunshine grows crops faster, it doubles or triples water demands. The extra heat combined with dry winds depletes soil moisture quickly and creates the need for higher irrigation volume and frequency.

Weather, Water and Waste

In normal years, California produces over half of the vegetables and fruit consumed in the U.S. The state irrigates 9.6 million acres using roughly 34 million acre-feet of water from lakes, reservoirs and rivers or pumped from groundwater.

California, like the Rocky Mountain States, is experiencing its worst drought in history.[24] The Central Valley Authority that distributes irrigation water announced a zero allocation to many crop regions. The Bureau of Reclamation estimated that one million acres would be put out of production and another two million acres would grow less food than normal. He called the situation grim.[25] By 2020, California will face a shortfall of fresh water as great as the amount that all of its cities and towns together consume in 2012.

California agriculture consumes 81% of the water in the state while providing only 3% of the state's revenue. Agriculture fights with cities for access to water that is diverted mostly from the northeastern part of the state in the Sierra Nevada Mountains to the fertile deserts of the Central Valley and southern California.

Interior Secretary Ken Salazar announced in July 2009 that $40 million in federal stimulus funds would be spent to pull more water from depleted underground aquifers in drought-stricken California. This political gambit came despite clear evidence that the well drilling will degrade the quality of water delivered to millions of residents. The stimulus money dug 50 new wells and installed temporary pipes and pumps to move water to crops and orchards.[26] While scientific consensus shows the state's aquifers are quickly being drawn down, no rules govern how much water can be pumped.

Fresno's water level has dropped from 20 feet below ground in 1950 to 120 feet below the surface in 2003. When water tables drop, springs that start rivers, along with lakes, reservoirs and wells go dry. Drawing groundwater creates subsidence and sinkholes, which break bridges, building, roads and canals.

An estimated 40 million people died during the 20th century's drought-induced famines, and the 21st century's toll could be higher. Over 30 million people depend on Colorado River water, which was divided among several states during plentiful rain years. As less rain and snow fall, water fights are becoming increasing fierce. The Colorado River now is practically dry by the time it reaches Mexico.

The Heartland is on track to crash their groundwater at about the same time California groundwater goes dry. Midwest farmers mine over five trillion gallons of fossil water a year from the Heartland's aquifer that covers eight High Planes states from North Dakota to Texas. The Ogallala aquifer contains water laid down millennia ago and is not replenished by annual rains. Children of Midwestern farmers will not inherit the valuable farmlands they expect but only dry prairie land that will be practically worthless. They will not be able to grow crops because all the irrigation water was mined by prior generations. Major cities may face migration if new water sources cannot be found.

Lake Powell 80% dry and Colorado River 90% dry in Mexico

Over-drafting causes groundwater tables to drop so that hydrology defeats farmers as the cost of pumping becomes prohibitive. As wells go deeper, water often becomes unfit for crop production. Dissolved salts create brine water that kills land plants. Deep groundwater often contains too many heavy metals such as mercury, lead and arsenic. Deep water in California's Central Valley contains too much boron, which destroys not just the crop but also the cropland.[27]

Much of the world's grain is produced with fossil aquifer water that is non-rechargeable. Overdrafting threatens aquifer crashes within the current generation. Many fossil aquifers have already crashed, which immediately stops food crop production.

Federal and state governments allow landowners with wells on their property to swap their underground water for higher-quality canal water. The brackish groundwater is then pumped into the aqueduct that supplies cities, farms and industry to the south with higher saline water. The saltier water will cause $18 million in economic damage a year for water recycling plants and appliances deterioration.

Drought also plagues the South and Eastern U.S. From 2008 through 2012 Texas farmers received minimal or no fall rain and lost their winter wheat crop. Farmers were not able to plant spring crops because there was no soil moisture.

Water conservation and regulation policy is problematic because Americans views groundwater as a property right. T. Boone Pickens' Mesa Water is buying up water rights to sell water to cities like Dallas.

Rational government policy would limit irrigation to sustainable yields. Instead, government policies in the U.S. and globally have encouraged maximizing short-term food production by subsidizing water including transport, delivery and the energy needed for pumping. Over-pumping at several times the sustainable yield has resulted in plunging water tables on every food-growing continent. Many aquifers are falling at 10 feet a year and several will crash before 2031.

Over 60% of India's grain production depends on irrigated land.[28] More than half of this irrigation water comes from groundwater aquifers that are being depleted at 300 times nature's replacement rate. India's 100 million farmers have drilled 21 million irrigation wells over the last 30 years and half of those wells and millions of shallower tube wells have already gone dry. In many areas, moving water for irrigation consumes 80% of the grid energy and power outages are very common.

The Gangotri Glacier in the Himalayas has provided 70% of the water for the Ganges River for centuries but is retreating 35 yards each year.[29] When the glacier melts, the Ganges will become a seasonal river depending on annual rains depriving 40% of India's irrigated cropland and nearly half a billion people of water.[30]

Climate Independent Foods

While India makes a disturbing case study, analysis in every crop-growing region shows similarities, including many U.S. farmers. Hotter temperatures force farmers to use more water. As water tables drop, the cost of pumping increases until wells are too deep to be economically feasible. Wells around the edge of the Ogallala aquifer in parts of New Mexico, Texas and Oklahoma have already run dry. People from towns like Happy Texas have had to migrate. When the Heartland produces no crops, what will our children do for food?

Waste and pollution

Modern agriculture disrupts and degrades its own ecosystems with massive waste streams that pollute the air, water and soil. Industrial agriculture accounts for accounts for 52% of the global anthropogenic methane emissions and 84% for nitrous oxide.[31] Field crops and soils act as a CO_2 sink, but the net flux is small. Crop cultivation, harvest, processing and transportation add millions of tons of CO_2 to the atmosphere each year.

The cost of nitrogen, (N) fertilizer has tripled recently because 90% of the cost comes from the natural gas used to synthesize it. Each U.S. corn crop removes more than 6 billion pounds of N, hay grown to feed livestock removes 7.4 billion pounds and wheat extracts another 2.4 billion pounds of N from soils.[32] The massive nutrient erosion severely pollutes fragile ecosystems, waterways, and well water. Soil erosion in the U.S. does about $45 billion in damage each year.[33]

Livestock are the single largest user of land. Meat production accounts for 70% of all agricultural land and 30% of the land surface of the planet.[34] Roughly one-third of the world's food grains go to feed livestock for meat and dairy products. Livestock are responsible for 18% of all greenhouse gases, more than all cars and SUVs combined.[35] Livestock contribute 37% of the methane and 65% of the nitrous oxide to the atmosphere, which are worse greenhouse gases than CO_2. A single cow—calf pair on a beef farm produces more gas emissions than a person driving 8,000 miles in a midsize car.[36]

Industrial agriculture is the leading source of water pollution in the U.S. The EPA National Water Quality Inventory notes that agricultural pollution is the source for about 48% of polluted surface water.[37] The

Weather, Water and Waste

EPA reports that 46% of America's rivers are too polluted for fishing, swimming, or aquatic life in 2012. Two-thirds of U.S. estuaries and bays are either moderately or severely degraded from eutrophication (nitrogen and phosphorus pollution). The Mississippi River carries an about 1.5 million metric tons of nitrogen pollution into the Gulf Mexico each year. The resulting hypoxic coastal dead zone in the Gulf each summer is about 8,500 square miles, larger than the state of New Jersey. A review study in *Nature* reported that the over 420 dead zones globally are growing at about 10% a decade.

Dead Zone at the Mouth of the Mississippi River

Row crops like corn and soy amplify erosion and pollution. A typical acre of corn raised with industrial methods in Michigan produces about 150 bushels of corn. Since corn contains about 23% protein, the acre produces about 200 pounds of protein. Input requirements include approximately 5 gallons of diesel fuel, 28,000 genetically engineered seeds; 150 pounds of nitrogen, 55 pounds of phosphorus, and 85 pounds of potassium fertilizers; a gallon of herbicide, and a gallon of pesticide/fungicide.[38]

Row crops increase soil erosion from both wind and water because there is nothing to slow the wind or water as it blows or flows down the rows. Soil erosion typically occurs at the top of the soil where fertilizers, pesticides and herbicides were applied, which causes pollution to the air and creeks, streams, wetlands and lakes.

Farmers apply huge amounts of fertilizer but about half is wasted as it erodes and flows into wetlands. Pesticide waste is even worse since pests consume only about 2% of the applied pesticides. The remainder pollutes local and regional ecosystems. The U.S.GS reported the 2008 spring run-off carried 817,000 tons of nitrogen, roughly 45% above normal, and 85,000 tons of phosphorous about 85% percent above normal to the Gulf. [39]

An Iowa Department of Natural Resources report indicated 274 Iowa waterways were seriously polluted. Fertilizer run-off causes such a problem that Iowa boasts the largest and most expensive nitrate removal plant in the world.[40] Downstream states such as Oklahoma experience even more agricultural pollution, where 94% of lakes fail federal water quality standards and 75% of the rivers are polluted.[41]

The most commonly used pesticide, organophosphates cause delayed-onset toxicity to nerve cells, which is often irreversible. The EPA lists organophosphates as very highly acutely toxic to bees, wildlife, and humans. Several studies have shown persistent deficits in cognitive function in workers exposed to pesticides.[42] Evidence suggests that even at very low levels of pesticide exposure in foods or fields, have adverse effects in the neurobehavioral development of fetuses and children.

Summary

The combined action of increasing severe weather, water and waste challenges place our food supply at serious risk. Health issues related to industrial foods create severe drag on people, their families and society. Added to these issues, the overconsumption of fossil resources will crash our food supply and leave the next generation with insufficient natural resources to produce food.

Chapter 2. What will our Children do for Food?

We need a food revolution to leave sufficient natural resources and clean ecosystems for our children.

s

Weather, water and waste independence are insufficient to assure sustainable and affordable food production. Our current fossil food supply path massively consumes non-renewable resources, wastes them and then allows those natural resources to severely pollute air, water and soil.

When our children are hungry and need food, they will find the fossil resources needed to produce industrial foods are extinct and their ecosystems hopelessly polluted. What will our children do for food? They will need a fossil-free food supply.

Prince Charles has advocated sustainable agriculture with less dependence on fossil resources for over 30 years. He shared his insights in a keynote address at the Georgetown University Future of Food Conference. He conveyed his motivation as:

> I have no intention of being confronted by my grandchildren, demanding to know why on earth we didn't do something about the many problems that existed when we knew what was going wrong. The threat of that question, the responsibility of it is precisely why I have gone on challenging the assumptions of our day. And I would urge you, if I may, to do the same because we need to face up to asking whether how we produce our food is actually fit for purpose in the very challenging circumstances of the twenty-first century.

Climate Independent Foods

> We must take care of the earth that sustains us because if we don't do that; if we do not work within nature's system, then nature will fail to be the durable continuously sustaining force she has always been. Only by safeguarding nature's resilience, can we hope to have a resilient form of food production and insure food security in the long term. This, then, is the challenge facing us.[43]

Prince Charles provided evidence of scarcity and unsustainability for each of the fossil resources required to produce industrial foods. Prince Charles advocates for genuinely sustainable agriculture for the long term that replenishes soil and water, is not dependent upon the use of chemical pesticides, fungicides and insecticides nor artificial fertilizers, growth promoters or GE monocultures. He noted that we must reduce the use of those substances that are dangerous and harmful, not only to human health but to the health of those natural systems such as the oceans, forests and wetlands.

Prince Charles recognizes that food production concentrated in large farms distant from consumers puts our food supply in jeopardy. A single weather or geological event can destroy large food stocks or growing regions. In addition, with increasing fuel scarcity and cost, long distance food transportation will not be possible. He advocates a more resilient system were many smaller growers produce food close to consumers. He notes "...strengthening small farm production could be a major force in preserving the traditional knowledge and biodiversity that we lose at our peril."[44]

Many other voices agree that industrial agriculture is unsustainable and argue for distributed, small organic farm production. While those actions will act to conserve resources, they are equally prone to threat from climate change, water scarcity and pollution. A better solution lies in growing crops independent of the factors that make modern agriculture, industrial or organic, unsustainable. The architecture for a new sustainable food system needs to make corrections for the design flaws in modern food production.

What will our Children do for Food?

People

Our food supply system should provide healthy, clean and nutritious food for people. Industrial farmers use predominately GE seeds to produce monocultures that are refined to foods high in fat, sugar, salt and preservatives, but low in nutrients per calorie. Crops grown in nutrient deficient soil suffer from hidden hunger, and may have up to 75% less of the deficient nutrient in foods compared with fertile soil.[45]

Nutrient deficient plants may cannibalize their cell walls and storage nutrients to insure survival. Nutrient deficiencies are often visible in the field, but may not be apparent in the produce. Produce with nutrient deficiencies transfers hidden hunger to human consumers, which expresses in the many maladies and diseases caused by nutrient deficiencies.

Hidden hunger from nutrient deficiencies imposes a huge toll on society, according to the UN World Health Organization, (WHO).

- Vitamin and mineral deficiencies account for 10% of the global health burden – second only to clean water.
- Children and adults with micronutrient deficiencies suffer impaired development, disease and premature death.
- Over 2 million children die unnecessarily each year because they lack vitamin A, zinc or other nutrients.
- Over 18 million babies are born mentally impaired due to iodine deficiency each year.
- Iron deficiency undermines the health and energy of 40% of women in the developing world. Severe anemia kills more than 50,000 women a year during childbirth.[46]

Hidden Hunger

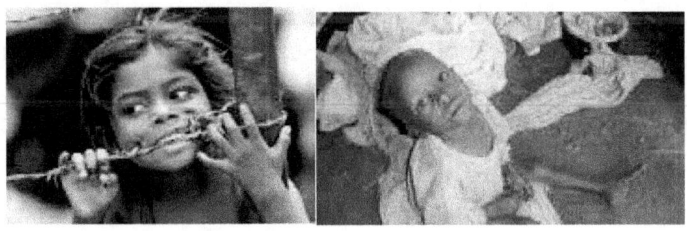

Climate Independent Foods

Scientists use the term empty calories to describe how nutrient dilution from industrial farming diminishes the nutrients in each bite. Nutrient dilution occurs because farmers maximize yield weight, not food quality or nutritional density. Yield weight increases with GE foods, but most the extra weight comes from water. Nutrient dilution also depletes modern foods of color, aroma, taste, texture and nutrient density. GE monocultures jeopardize our entire food supply because a single pest vector can destroy the entire crop. GE monocultures also cheat consumers of nutralence, nutrient density, and diversity.

Our human bodies evolved over eons to eat nutritionally rich and diverse natural foods; not GE monocultures loaded with preservatives and chemical residues. GE crops require massive amounts agricultural chemicals and poisons because they are genetically engineered to optimize yield, not plant vitality. GE crops are extremely vulnerable to a broad spectrum of weeds and pests, so they require significantly more fertilizer and chemical poisons, some of which stays as residue in refined foods and fresh produce. Recent scientific evidence suggests that despite Monsanto's promise of GE food safety, genetic crop manipulation causes changes in human body chemistry.

Chemicals in fertilizers and pesticides have been linked to ADHD, autism, cancer, Lou Gehrig's disease, and other illnesses. Three major studies published in May 2011 in *Environmental Health Perspectives* shows children exposed to pesticides in the womb are more likely to have measurable problems with intelligence, memory, and attention beginning at 12 months and continuing through early childhood.[47]

These studies link prenatal pesticide exposure (measured in the urine of mothers-to-be) to significantly lower IQ in children by age 9. The research teams, from Mount Sinai School of Medicine, Columbia University's Mailman School of Public Health and the school of public health at the University of California, Berkeley, all conclude that pesticide exposure during pregnancy could negatively affect brain development.

What will our Children do for Food?

Animal studies have demonstrated that organophosphates, (OP) scramble brain function and behavior in baby rats. Two studies in 2010 found that children exposed to higher levels of organophosphate pesticides than their peers were more likely to be diagnosed with attention deficit hyperactivity disorder, (ADHD).

One recent study followed hundreds of mostly Latino mothers and children in California's Salinas Valley, a center of commercial agriculture. Many of the women were farmworkers, or had family members who worked on farms. When the women were pregnant, the researchers tested their urine for several chemical by-products of organophosphates -- a standard means of gauging exposure. The mothers with the highest levels of by-products, known as metabolites, had children whose IQs at age 7 were seven points lower, on the 100-point scale), than the children whose mothers had the lowest levels of exposure.

Americans, like people in all food growing regions, are exposed to air pollution, partially due to agriculture. The American Lung Association's 2011 State of the Air reports that 155 million Americans, just over half the nation's population, live in areas where air pollution levels are often dangerous to breathe.[48]

Leonardo Trasande and colleagues at the Mount Sinai School of Medicine estimated the annual cost of environmentally mediated diseases in U.S. children to be $76.6 billion in 2008.[49] Since roughly half of air pollution and a majority of water pollution come from industrial agriculture, the agriculture creates a cost to our children of at least $33 billion. Of course, these medical costs ignore the loss of life quality and family disruptions caused by childhood cancer, asthma, intellectual disability, autism, and attention deficit disorder. No similar studies have examined the elderly but the costs would probably be higher since the elderly are a larger group and have substantially longer exposure to environmental pollution.

New food production should avoid hidden hunger and empty calories. It should avoid GE foods and agricultural poisons. Growing crops should not pollute air, water or soil and not expose people to pesticide residues.

Climate Independent Foods

Producers

Fossil foods adversely affect growers too. Genetic engineers trick plants to transfer their limited energy from roots to fruit, which create more crop by weight, but at a steep price for consumers. Much of the additional weight comes as water weight, which cheats consumers with additional nutrient dilution.

GE plants cannot compete with natural weeds, so farmers have to apply additional agricultural poisons – pesticides, herbicides and fungicides. The poisons, combined with fertilizers and other agricultural chemicals, put farmer and farm family health at risk. Millions of farm animals die each year due to exposure to farm poisons. A recent study by the Union of Concerned Scientists revealed that despite of Monsanto's claims that GE crops use fewer agricultural poisons, producers are actually using more pesticides and herbicides.[50]

Farmers must pay monopolistic prices for their GE seeds and cannot save their seeds, as farmers have for eons. Farmers often have to pay more for their seeds each year. GE crops need protection from natural plants, which add to farmer cost and labor. Shorter roots require farmers to apply additional irrigation to retain soil moisture in the narrow root zone. Shorter roots limit the plant's reach for nutrients, so farmers must apply significantly more chemical fertilizers.

Producers are exposed to hazardous agricultural chemicals

New crop production should protect the health of farmers, farm families and their neighbors.

What will our Children do for Food?

Planet

Soil degradation and loss operate in a slow insidious manner that erodes farmers' ability to grow nutritious crops. Plant pathologist Don Huber, professor emeritus from Purdue notes that crops usually get enough phosphorus, potassium and other common minerals to grow, but often cannot draw sufficient micronutrients from the soil to fend off diseases.[51] Such nutrients include the metals manganese, copper, zinc, iron and boron. Crops deficient in micronutrients pass their hidden hunger on to animal or human consumers.

In order to maximize short-term production, industrial agriculture manipulates the natural ecosystem and disrupts, degrades and displaces precious soils and nutrients. Industrial agriculture accelerates erosion around 500 times faster than natural processes.[52] Soil carried away by erosion contains about three times more nutrients than are left on the remaining soil.[53] Bruce Wilkinson estimates that global erosion occurs at a rate of about 75 billion tons a year.[54] Moving these amounts of rock and soil would fill the Grand Canyon in Arizona in a generation.

Industrial agriculture erodes 1.8 billion tons of soil from U.S. cropland each year.[55] According to the EPA, agricultural pollution leaves over 52% of our waterways unfit for human recreation.[56] In some agricultural states, pollution makes over 90% of the waterways unfit for fishing.

Cropland erosion

Fossil foods place farmers and their rural neighbor's health in jeopardy from environmental pollution. Erosion not only removes topsoil, organics and nutrients, but also carries sediment to waterways along with agricultural poisons. Most crops absorb only about 5% of the pesticides applied. Most of the residual enters wetlands, waterways and groundwater.

Industrial agriculture's magical promises have turned into a nightmare in India. Fertilizers and pesticides have as so polluted farmlands that thousands are dying. Trains from the city of Chotia Khurd in northern India are now called cancer trains because so many people in the farming villages must go to the city for cancer treatments.[57]

Agricultural Poison Pollution

Over eight million Indian farmers quit farming during the 1990s due to rising crop input prices – seeds, fuels, fertilizers and chemicals – that created escalating farmer debt. In the decade ending in 2007, over 183,000 farmers in India committed suicide because their farms could no longer provide for their families.[58] Government sources note that farmer suicides are substantially under-reported. Additional millions of farmers and family members have died or become disabled due to agricultural poisons.

New food production should end soil erosion and pollution. Ideally, growing crops should repair damaged fields and local environments.

Rather than continue on a path of overconsumption of natural resources and environmental destruction, we might consider adopting a Food Bill of Rights. These rights benefit people, producers and our planet.

Chapter 3. What is the Food Bill of Rights?

> The test of our progress is not whether we add more to the abundance of those who have much; it is whether we provide enough for those who have too little.
>
> *– Franklin D. Roosevelt*

The new architecture that will transform our food supply from insecure and unsustainable to secure and sustainable requires at least ten design constraints. The Food Bill of Rights and Protections offers liberty to consumers and growers and benefits to stressed ecosystems, Table 1.

1. Food security. Access to affordable good food represents a fundamental right and creates social equity. When people lack access to good food, they create a substantial drag on all society. When all people have access to food, everyone can contribute to improving society, our food, art, business and our ecosystems.

People should have access to good food or the inputs for growing their own food. Over half our global population, 3.5 billion people are food insecure and often hungry.[59] Many of their children and elders are malnourished. A child in Africa dies of malnutrition every 5 seconds.[60]

Poor Americans lack food freedom. They must feed their children cheap, nutrient deficient foods that cause obesity, fatigue, dull brains and diabetes. The poor have no choice because healthy foods are often unavailable or unaffordable.

Table 1. Food Bill of Food Rights for People, Producers and our Planet

Food rights	Description
1. Food security – Food democracy	All people have access to sustainable, affordable, and delicious food.
2. Healthy choices	Natural and clean and low in fat and cholesterol. Cures rather than causes obesity, diabetes, and other diseases.
3. High nutralence	High micronutrient availability and density. No empty calories and delivers high levels of essential nutrients in every bite.
4. Excellent sensory appeal	Superior color, aroma, texture, and taste compared with fossil foods.
5. Creates local jobs	Offers good local jobs with a living wage.
6. Preserves natural resources	Fossil free and avoids extraction, which preserves natural resources for our children.
7. Fresh and local	Meets our 50/50 goal where 50% of food grows within 50 miles of consumers.
8. Naturally clean and biodiverse	Millions of natural species grow, which avoids the need for genetic monocultures.
9. Cleans and repairs ecosystems	Cleans and regenerates ecosystems so growers can leave every field better than they found it.
10. Climate independence	Reliably produces food independent of climate, weather, geography, or politics.

What is the Food Bill of Rights?

USDA food policy has kept food costs relatively low, at about 9% of consumers' income. Unfortunately, the unintended consequence of cheap food is a doubling of medical costs that now approach 18% of consumers' incomes. Some economists predict health costs will exceed family income by 2025.[61]

The structure of our food supply should be a top priority for health because available food has significant impact on medical costs. Food security also plays important roles in national defense, security and education. Politicians and policy leaders ignore the strategic value of our food supply – to our peril.

Most people cannot grow their own food because the prices of food inputs continue to escalate. Many municipalities have passed laws forbidding the use of some fertilizers and poison that damage local ecosystems. Unsurprisingly, the largest polluters, agribusinesses, are exempt from these laws.

New food architecture should assure food security for all.

2. Healthy choices are not available to most consumers today because freedom foods are not yet widely available. Most industrial foods create rather than prevent health problems.

Modern foods lack informative labels in the U.S.. The European Union considers food labels the most important tool for ensuring the freedom of choice. The EU gives this freedom to know and freedom to choose by law. Genetically engineered, (GE) food products must clearly state the presence of GE food on the label. Clear labels give every consumer an opportunity to make an informed decision.

Freedom to choose non-GE foods does not currently exist in the U.S. because the FDA does not yet require labels for GE crops. GE monocultures dominate the U.S. food supply, which puts the entire food supply in jeopardy. Over 90% of basic food ingredients in the U.S. come from GE crops: soy, corn, canola, and sugar beets.

Many consumers, especially children, must eat cheap foods high in fat, salt and calories but low on nutrition because they cannot afford healthy foods. Consumers deserve affordable healthy foods that promote strong minds and bodies. Foods should prevent

Climate Independent Foods

rather than cause obesity, diabetes, heart disease, cancers, and the other Western diseases.

Consumers expect clean, natural foods. Fossil food production requires heavy chemical fertilizer and pesticide applications to avoid natural competitors and to maximize yield weight. Most of the processed food grains in the U.S. contain unnatural, GE material as well as synthetic preservatives and pigments. Fresh and packaged foods also contain chemical, pharmaceutical and pesticide residues.

GE crops require farmers to apply huge amounts of agricultural chemicals and poisons, which imposes substantial health risk to farmers, their families and their farm animals. Aerial spraying drifts far from fields and flows into wetlands and waterways. Exposure to polluted ecosystems creates prolonged health risk for rural communities, diminishing their quality of life, and the value of their property.

Industrial farmers also experience substantial physical and health risk from working with heavy equipment, hard physical labor, and fatigue. Modern agriculture competes with fishing and mining for the industries with the highest rates of disabilities and death.

A transformational food system should provide for the health of people, producers and our planet.

3. High nutralence. Our food supply should liberate our children from hidden hunger and the silent but costly nutrient deficiencies stemming from nutrient dilution. Fossil foods create nutrient dilution because the crops systemically extract soil micronutrients that farmers do not replace. (Farmers do replace the macronutrients in order to maximize yield.) Consumers deserve to know when nutrient dilution occurs in produce or processed foods. Current food labeling provides far too little information on nutrient availability, density, quality and diversity.

The Centers for Disease Control, (CDC) reported that one out of three American children born after the year 2000 will contract diabetes – predominantly due to a poor diet of nutrient-deficient calories.[62] Over 40% of women are likely to contract diabetes. The plague of obesity and diabetes creates havoc on our educational system and creates immense drag on our health system. The CDC estimate the total costs from diabetes at $174 billion annually.[63]

Consumers should have access to healthy foods with high nutralence that prevent rather than cause diabetes and other Western diseases.

4. Excellent sensory appeal. Nutrient dilution and hidden hunger in crops causes empty calories, which deliver fewer nutrients per bite. Low nutralence diminishes sensory appeal, color, taste, aroma and texture. A modern field tomato has poor aroma, taste, color and texture due to industrial food cultivation and harvesting practices.

Consumers should have access to fresh whole foods with excellent sensory appeal, especially taste.

5. Local jobs. Growing food should produce local jobs. Local jobs are practically impossible with industrial agriculture. Concentration of power in big agribusinesses siphons off the wealth from family farms and concentrates money spent on food in the pockets of a few. Family farmers are unable to compete with huge producers who monopolize access for the fossil resources required to produce food. Large agribusinesses employ legions of poorly paid, unskilled labors that have sparse access to good education, housing, healthcare and healthy food.

Farm subsidies magnify inequity and waste billions. Over 70% of the $170 billion in farm subsidies over past 15 years supported the production of just five crops: corn, wheat, cotton, rice and soybean. Just four of those same favored five: corn, wheat, cotton, and soybean accounted for over 70% of the $25 billion in crop insurance over 15 years.[64] Widespread drought and heat caused crop losses in 2012 will elevate crop insurance costs.

From 1995-2009, the largest and wealthiest top 10% of farm program recipients received 74% of all farm subsidies with an average total payment over 15 years of $445,127 per recipient. The bottom 80% of farmers received an average total payment of just $8,682 per recipient.[65] Large agribusinesses and wealthy farmers game the system to gain the agricultural subsidies put in place to support small family farms. They use the wealth given by the government to buy land and other resources from smaller farmers, many of whom do not receive subsidies.

Subsidy policy should make agriculture cleaner, healthier and more sustainable. Farm policies should support social justice. Current U.S. agricultural subsidies amplify social inequity, enrich those who are already wealthy, and promote only a few crops.

Canada, Mexico and other close allies have outstanding lawsuits against U.S. subsidies with the World Trade Organization because these subsidies substantially depress the real price of food grains. British International Development Secretary Douglas Alexander said "It's unacceptable that rich countries subsidize farming at $1 billion a day, costing poor farmers in developing countries $100 billion a year in lost income."[66] Subsidies hide the true cost of food, create incentives for unsustainable farming practices and bloat the financial deficit. We need to transform subsidies so they motivate production of environmentally sustainable crops that repair ecosystems.[67]

The World Trade Organization ruled that U.S. cotton subsidies, contravene international rules. The U.S. government "settled" the suit in 2010 by agreeing to subsidize Brazil's cotton farmers over half-billion dollars over the next several years.[68] WTO subsidy lawsuits from Canada, Mexico and several African countries remain unresolved.

Growers and the U.S. economy will benefit from a food system that transforms crop subsidies and creates good local jobs that provide a living wage. Consumers will benefit too with healthier food choices.

6. Preserves natural resources. The only way to assure food security and sustainable production is to drastically cut or eliminate the consumption of fossil resources. A food supply based on fossil resources imposes five fatal errors on societies; constantly rising prices, freshwater scarcity, soil erosion and degradation, severe environmental insults and eventually crop failure.[69]

Food prices escalate in lockstep with the price of oil and other fossil resources because these resources are necessary for production. The price of a single fertilizer, phosphorous, rose 700% over a recent 14-month period. As natural resource scarcity increases, food costs will rise even faster because speculators will hoard resource assets. Speculators have already disrupted land, oil, water, fertilizer and commodity markets.

The increasing heat from global warming drives soil moisture from soils and plants, multiplying the need for freshwater. Recent cropland expansion has converted deserts to cropland with irrigation. More sunshine is great for crops but desert farming consumes 200 to 500% more freshwater. Desert farming wastes huge amounts of water that percolate quickly through sandy soil.

Countries, states, cities and farmers have been fighting over water for decades. As global warming intensifies droughts and causes more water loss from evaporation, water conflicts will intensify. Vandana Shiva in *Water Wars: Privatization, Pollution and Profit* predicts the water wars of the 21st C may surpass the oil wars of the 20th C.

Farm operators now lease over half the cropland in the U.S.[70] Non-owners are motivated to maximize short-term profitability and have less motivation than owners to use sustainable agricultural practices. Systemic extraction, waste, and pollution systemically degrade ecosystems until they become unproductive and must be abandoned.

Industrial agriculture promotes erosion by plowing fields with heavy equipment, intensive cultivation, and application of harsh fertilizers and agricultural chemicals. Growing crops constantly disrupts and compacts topsoil while extracting nutrients and humus, which amplifies erosion and wears out the soil. As fields deteriorate, farmers react by applying more water, fertilizers and pesticides. These actions not only increase costs but also make run-off more toxic.

China's farmers must abandon over a million acres of degraded cropland each year. The resulting dust storms are worse the U.S. Dust Bowl in the 1930s. China's air pollution caused over 20 million people to suffer respiratory illnesses in 2007. The country's health ministry demanded that the World Bank remove mortality calculations from a report on the country's air and water pollution because the numbers could trigger social unrest.[71]

Climate Independent Foods

When soil, oil, freshwater, phosphorus or any other nonrenewable input becomes unaffordable or extinct locally, the food supply crashes. Food supplies have crashed in many countries already. In the Mid-East, insufficient freshwater ended food production in many regions. In parts of Central America, Africa and China, erosion and expanding deserts have destroyed cropland.

The serious threat hidden in the fossil foods is that when just one of the fossil inputs becomes unavailable, the entire food supply fails.

Sustainable food production requires freedom from fossil resource consumption.

7. Fresh and local. Consumers deserve access to fresh foods grown locally, which avoid long supply chains that rob foods of nutrients and taste. Local foods can reduce up to 60% of food costs by eliminating most the refining, preservatives, packaging, storage, and spoilage. Local foods avoid the extensive energy and pollution caused by long distance transportation.

Industrial food production occurs far from consumer concentrations in cities. The cost of land plus the dust and need to use agricultural poisons make industrial food production inappropriate for cities.

Local food production is largely impossible, except for the few small producers for farmers' markets. The USDA farm policies have benefited rich agribusinesses but largely decimated local family farms. The 322,000 principal U.S. farm operators, only 0.001% of the U.S. population, on 322,000 farms, produce 90% of all foods and fibers consumed in the U.S., plus another 11% for export.[72]

Government subsidy policies designed to support family farmers instead spawned a rich oligarchy at the top U.S. agricultural production. These few largest and wealthiest agribusiness scions orchestrate Congress to award continually more transfer payments in the form of subsidies, tax relief, trade barriers, foreign aid, and other incentives to enrich themselves. In contrast to political rhetoric, small farmers realize negligible benefits from government farm policies, and subsidies.

What is the Food Bill of Rights?

Food production should meet the 50/50 goals where 50% of food consumed is produced locally, within 50 miles of consumers. The 50/50 local production could save five times more transportation fuel than the entire 48 million acres of corn grown for ethanol.[73]

8. Naturally clean and biodiverse foods enhance food security and quality while providing nutrient diversity to consumers. Nature has used the natural biodiversity strategy for eons to assure sustained plant production. A biodiverse food system creates resilience than can withstand a wide spectrum of stressors or invaders.

GE monocultures are not only bad for human and animal health but put the entire food supply at risk. A single pest such as a fungus, mold, mildew, insect, worm or weevil can devastate a monoculture in a single growing season.

A natural biodiversity strategy gives crops resilience and growers the freedom to choose among thousands of desirable crops.

9. Cleans and repairs ecosystems. Society should be freed of waste and pollution. Soil pollution costs U.S. citizens over $45 billion annually – about 50% more than the U.S. spends yearly on health research.[74] Yale professor Robert Mendelsohn and Nicholas Muller estimate air pollution (from all sources in the U.S.) damages range from $75 - $280 billion annually.[75]

Farmers do not pay for their environmental destruction costs. Current social policy imposes no taxes on extraction, waste or pollution. Society pays, or more precisely, our children will bear these costs. The current food system will leave our legacy – severely degraded ecosystems barren of valuable natural resources. The amber waves of grain on our plains will be gone, replaced by dry and dusty deserts.

An unfortunate unintended consequence of fossil foods has been the public and environmental health impacts from soil erosion. High commodity prices and crop subsidies motivate farmers to plant crops on highly erodible wetlands and hillsides. Research at Iowa State University provides evidence that erosion loss in some regions occurs at 50 to 500 times faster than topsoil forms. Soil erosion occurs at levels far beyond government estimates.[76] The Iowa Daily Erosion Project reported that a single 2007 spring storm created an average loss between 20 and 40 tons per acre across 14 Iowa townships.[77]

Some farmers try to practice conservation agriculture, but industrial farming methods undermine their efforts. Crops continually extract macro and micronutrients, while cultivation disrupts and compacts the soil, amplifying erosion. Chemical fertilizers and poisons are toxic to soil microbes, which reduces soil fertility and further degrades soil structure. When agricultural production has depleted soil nutrients and degraded soil structure, farmers must abandon the cropland.

A food production system that cleans air and water will benefit everyone. A food production system that regenerates degraded soil will provide good food for millions.

10. Climate independence. The only way to assure food security with global climate chaos is to free farmers with climate independence. The food production system should produce reliably independent of climate, heat, cold, geography, or politics.

Climate independence does not mean food production in all weather. In closed systems, food production can be completely climate independent. In open and covered growing systems, climate independent foods produce 10 to 50 times more food during the growing season than terrestrial crops. Unlike land-based crops, these crops produce consistently, every season.

The next section explores how abundance methods grow freedom foods that support the Food Bill of Rights.

Chapter 4. What is Abundance?

Why should food producers continually extract declining fossil resources when growers can use plentiful resources that are surplus, cheap or free?

The design for an alternative food production system should produce food reliably, independent of weather. The process should avoid competition for natural resources with modern industrial agriculture. These resources include freshwater, fertile soil, fossil fuels, inorganic fertilizers as well as pesticides, herbicides and fungicides. Possibly the biggest challenge is the avoidance of waste and pollution.

Abundant agriculture provides a novel solution for sustainable and affordable food and energy, (SAFE) production that meets these challenges.[78] Abundance methods assure sustainability for at least seven generations. Growers recover, recycle and reuse bioenergy and nutrients in waste streams. Of course, safeguards are necessary to assure the demise of waste stream parasites and pathogens. Growers use simple solar heaters, UV light or other technology that has been effective for over 40 years old. Abundance methods cultivate algae that convert solar energy and free or low cost inputs into hydrocarbon-rich botanical biomass.

Climate Independent Foods

Abundance methods mimic nature's oldest and most reliable food production system at the base of the food chain. Without human cultivation, algae already synthesize roughly 0.8 x 10^{11} tons of organic matter daily, constituting about 40% of the total fresh organic matter grown on our planet.[79] Each day, algae supply 70% of the world's oxygen, more than all the forests and fields combined. Fortunately, algae's many hungry consumers eat most the new biomass daily or the earth would be covered in algae.

One hundred times more animals eat algae than any other food because it is so nutritious. Algae's tiny cell size, only 5 μ, make the plant and ideal nutrient delivery system for plants, animals and humans. Algae and algae infused crops offer 300 to 500% higher nutralence – high nutrient availability and density. The tiny yet nutralent cells are immediately bioavailable.

Abundance takes its name from the inputs, which are plentiful and often free, surplus or cheap – sunshine, CO_2 and wastewater. Abundance growers transform low cost inputs to cultivate valuable products. Abundance growers can clean polluted water and air while creating carbon neutral and water neutral food and energy. Every pound of algae sequesters two pounds of CO_2. The only thing released by abundance methods is pure oxygen.

Growers cultivate algae and other microorganisms in microfarms. Microfarms are flexible microcrop platforms that produce foods from plants lower on the food chain than modern industrial crops. Extensive research shows that foods low on the food chain, such as algae, provide superior nutrition for people, animals and plants than modern industrial foods higher on the food chain.

Algae, and the other microorganisms algae attract, grow protein and other nutrients 20 to 30 times more productively than land plants. This means a corn farmer would have to farm one acre for 30 years to produce as much protein as an abundance grower cultivates in one year. Growers can double the biomass daily allowing growers to harvest half the biomass every day the sun shines, year round.

Earthrise, an algae producer near the Salton Sea in southern California, has been cultivating algae for over 30 years in open raceway ponds. Earthrise produces about 500 tons annually of spirulina in 30 one-acre raceways that are about 12" deep. Spirulina is digestible and contains about 70% protein. Earthrise packages the spirulina as a healthy nutrient supplement. The company grows only for seven months a year, during maximum productivity. Some algae companies grow a winter cultivar.

Algae flour can substitute for corn, wheat, rice, soy or other food grain.[80] Algae oils offer a healthier alternative to vegetable or animal oils because they are lower in fat and cholesterol. Algae oils are significantly higher in micronutrients, vitamins and antioxidants than vegetable or animal oils and deliver omega-3 fatty acids.

Algae offer extraordinary biodiversity. Only about 35,000 algae species have been characterized. Experts estimate the total number of marine and terrestrial algae species to be in excess of 10 million. Each algae species produces a unique combination of protein, lipids, carbohydrates and specialty compounds. Algae culture collections available at several universities and institutes offer searchable algae lists for targeted compounds.

Algae distribution

Algae thrive all over the planet, including under the North Pole, under glacial ice in Tibet and under ice sheets in the Alps. Forests of kelp, a macroalgae or seaweed grow under the North Pole. Millions of tons of microalgae grow in both poles and form the base of the food chain for krill and other voracious algae feeders.

Algae crusts form a matrix in the hottest deserts that holds soil and minimizes erosion. Algae crusts provide the structure and nutrients that support plant germination and growth in desert environments. Therefore, several algae species can be found locally or adapted to nearly any climate.

A handful of local dirt or cup of water may contain over a trillion cells and more than 100 species of indigenous algae. Indigenous algae typically out produce species from distant algae collections.[81] Local algae evolved over eons and adapted to the local microclimate.

Climate Independent Foods

Growers can bioprospect local algae species to find robust varieties that can be grown for protein or targeted compounds for food, feed, fabrics, fertilizers, biofuels, nutraceuticals, cosmeceuticals, vaccines, pharmaceuticals or medicines.

Weather

Abundance growers cultivate food nearly independent of climate or weather. Algae have no true growing season, although most species hit maximum productivity from spring through autumn. Cold weather species grow in the winter, although many slow or go dormant in extremely cold weather.

Covered or closed microfarms allow growers to produce algae food in practically any climate, altitude, latitude or geography. Growers in extreme locations may need to supplement high-efficiency LED lights when sunshine is insufficient.

Open microfarms are modestly sensitive to weather. An acre raceway that typically produces 150 pounds of biomass a day may produce only 40% during a storm or on cloudy days. Unlike land plants, when conditions are not ideal, algae simply rest. As soon as sufficient sunlight appears, production returns to normal. Since growers can harvest year-round, stormy and cloudy days have only modest impact on total production. Extreme sunlight can be problematic with algae growth. Some growers diffuse sunshine to maximize production.

Microfarms in a controlled environment such as a greenhouse with optional LED lights are weather independent. These growers use sunlight when available and LED lights when solar energy is insufficient. Several microfarm vertical designs capture solar energy in the top culture and use LED lights to augment sunlight on lower layers. Some growers use highly efficient LED lights to extend biomass growth into the night. Several algae companies are experimenting with 24/7 algae production with LED lights.

Cold climate growers may grow cold tolerant algae in the winter. In Canada, Sweden and other cold regions, growers add heat to accelerate biomass growth. Heat and light energy may come from renewable energy such as solar, wind, waves or geothermal.

What is Abundance?

Water

Algae thrive with net zero freshwater. Algae cultivation can use non-potable water sources including brine, agricultural, municipal or industrial wastewater or ocean water. These water sources typically contain too much salinity or other pollutants for terrestrial crops. Wastewater must be cleaned with UV light, solar heaters or other tools to remove bacteria and pathogens before growing algae foods.

Land plants evolved from algae 500 million years ago. The first land plants probably used algae crusts for support as they developed their tiny root system. Roots severely limited terrestrial plants because roots made them immobile and dependent on sufficient moisture and nutrient availability within the rhizome.

Land plants die quickly with insufficient moisture because water carries the nutrients needed for cellular metabolism. If too little of any of the 24 essential nutrients are unavailable, land plants die or fail to produce seeds, so they cannot reproduce.

Algae grow to the limit of nutrients. If insufficient water or nutrients are available, algae take a rest and simply enter dormancy. When good growing conditions return, algae resume normal growth.

We have cracked open rocks thousands of years old to find algae cells on bioprospecting trips at Arizona State University. When cultured, these old cells begin their high velocity reproduction. Algae's dormancy strategy allowed these plants to sustain growth for 3.7 billion years in all types of extreme environments.

Land plants die in the presence of salt due to a plumbing problem. Salt ions in solution are too large for root absorption. The clog in the roots blocks water flow and causes the plant to starve. Algae evolved in ancient oceans, which were extremely saline. Algae have no roots and absorb nutrients directly into the cell, independent of the in-situ saline concentration.

Brine water makes up half the groundwater stored on the planet. Therefore, brine water alone provides a sustainable cultivation medium for algae foods. Brine water typically has lower salinity than ocean water. When crude oil pumps remove oil from the ground, they often recover 10 to 20 barrels of brine water for every barrel of oil. Today, this nutrient rich water is pumped deep into the ground to create pressure this is the removal of additional crude oil. Soon, algapreneurs will cultivate algae that remove the nutrients before the brine water is returned to the well. In many areas, brine water lies only a few feet under the ground and can be recovered by foot pumps.

In most cities, wastewater creates a huge cost. Algae can transform those costs to positive revenue; creating valuable biomass while cleaning the wastewater. The oldest algae application in the U.S. is wastewater treatment.

Many freshwater algae species can be trained to grow effectively in brine or wastewater. Many sea vegetables, edible seaweeds, also thrive in brine and saline wastewater.

Waste

Industrial agriculture produces massive waste and pollution. Abundance methods recover and repurpose bioenergy and nutrient waste into energetic foods with high nutralence. Abundance growers can clean polluted air, water and soil while they produce valuable algae biomass.

Currently, waste streams represent a huge cost for farmers and zoos. Zoos often have to pay more to dispose of the ZooPoo than they pay for animal feed. ZooPoo includes animal, botanical and trash wastes. The zoo experiences huge costs for handling, storing, inspecting, hauling and disposing of the ZooPoo. Currently, ZooPoo adds tons of biomass to landfills, which degrade the environment.

The amazing thing about ZooPoo is its value. Animal manure and botanical wastes retain roughly 60% of the bioenergy that was originally in the plant. Even more importantly in an economic sense, ZooPoo retains 80 to 90% of the original plant nutrients.

Abundance methods can recover and reuse the zoo waste stream and transform this huge cost to a profit center. The combined value of the reclaimed bioenergy and nutrients could cover the cost of animal care. A smart zoo will create an additional ZooPoo revenue stream with a world-class ecotourism exhibit that conveys how algae support sustainable green living.

Human, industrial and agricultural waste streams often contain heavy loads of fertilizers, pesticides, herbicides and fungicides that make them unacceptable for use on food crops. Tiny algae cells absorb individual elements from complex compounds such as pesticides, which detoxifies the agricultural poison. Current technology offers several algae solutions that detoxify waste streams.

Drugs create a fatal problem with manure as fertilizer. Organic food producers want to use animal manure as fertilizer but in many cases they cannot due to the pharmaceutical drug problem. For 60 years, meat producers have fed antibiotics to farm animals to increase their growth and prevent infections. Nearly 70% of all antibiotics produced in the U.S., nearly 25 million pounds a year are fed to cattle, pigs and poultry according to the Union of Concerned Scientists.

Feeding pharmaceuticals to animals sustains a growing demand for meat but it generates public health fears associated with the expanding presence of antibiotics in the food chain. Over 90% of the drugs in animals and humans end up being excreted either as urine or manure. Food crops absorb and concentrate antibiotics and other drugs when grown in soil fertilized with livestock manure.

Municipal and industrial waste streams often contain additional pathogens, pharmaceuticals and poisons. Algae offer solutions to detoxify these waste streams and to recover the nutrients.

Abundance provides air pollution solutions. Power, cement and manufacturing plants produce the heaviest load of carbon dioxide to the atmosphere. Several algae producers such as Carbon Capture Corporation are designing systems that sequester carbon from exhaust plumes. Current technology allows microfarms to capture only part of the waste stream because power plants operate 24/7 while algae grow during daylight hours. Since every ton of algae captures two tons of CO_2, waste plumes provide substantial economic benefit. Smaller microforms may locate near other carbon sources such as restaurants, cleaners or breweries.

Climate Independent Foods

Companies like Biovantage Resources are innovating with water pollution solutions. The use of algae to clean wastewater is the oldest algae application in the U.S. New algae technologies are making water reclamation more effective and efficient. Several companies have business models where their primary revenue stream comes not from the algae produced but recovery and resale of metals and nutrients. Biovantage Resources is currently installing and wastewater treatment system for a new gas recovery operation in North Dakota. The system uses solar energy augmented by LED lights to clean the water and produce algae products.

A dairy farmer in California may have to pay $0.35 per cow each day to dispose of liquid and solid wastes. The waste disposal cost for a 10,000 cow farm penalizes the farmer $35,000 a day. Farmers typically use the cheapest disposal method, which is burning or burying the waste.

Research at St. Cloud University in Minnesota has used algae to recover the residual energy from cow manure. This research suggests that the bioenergy value from each cow's manure is higher than the value of the milk. This calculation does not include the substantial savings from manure disposal.

Abundance products

Abundance methods can produce practically anything sourced today from land plants because land plants evolved from algae 500 million years ago. Most fossil fuels are made of algae or algae feeders fossilized from ancient oceans. Therefore, anything that can be made from fossil fuels can be made from algae. Algae biofuels burn without black smoke particulates because they missed the hundreds of millions of years in fossilization. Algae biodiesel burns cleanly, similar to vegetable oil.

What is Abundance?

Figure 4.1 Algae Products and Ecological Solutions

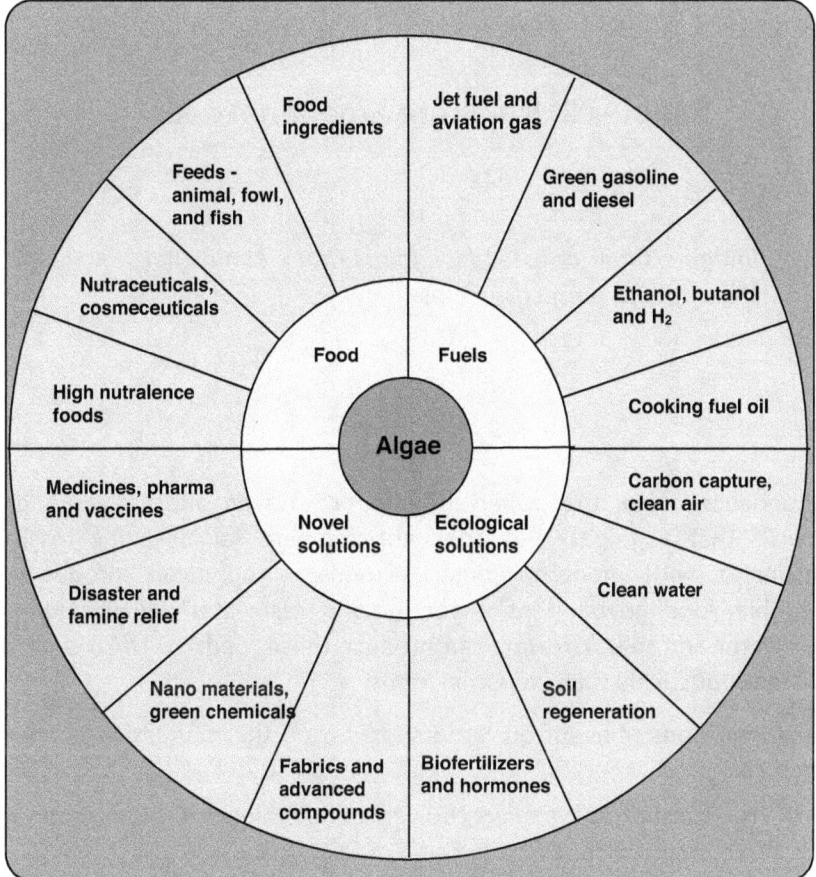

Other sources drill down into the many entrepreneurial opportunities with abundance production in each sector.[82] The focus here is on how abundance methods enable a new class of foods, freedom foods.

Chapter 5. What are Freedom Foods?

Imagine superior fossil-free foods that clean rather than pollute our ecosystems.

Abundance offers the opportunity to provide an alternative food source that augments but does not compete for declining fossil resources with modern industrial foods.[83] Abundance produces healthier food independent of weather and freshwater while cleaning air, water and soils. Growers can produce these foods 20 to 30 times more productively than industrial foods.

Freedom foods reinvent our food supply from the foundation of the food chain and:

- Free consumers for smart choices for healthier and tastier food.
- Free growers from consumption of fossil resources.
- Free ecosystems of waste and pollution and leave the environmental footprint of a butterfly.

Foods grown low on the food web require a fraction of the energy consumed by modern food. They are free of the resource consumption and waste caused by industrial crops because they are grown organically, with no or minimal fossil resources.

This alternative food supply provides splendid natural produce and does not compete with industrial agriculture because growers use abundance methods.[84] Growers recycle waste stream nutrients that are plentiful, affordable and will not run out.

What are Freedom Foods?

Figure 5.1 Freedom Foods

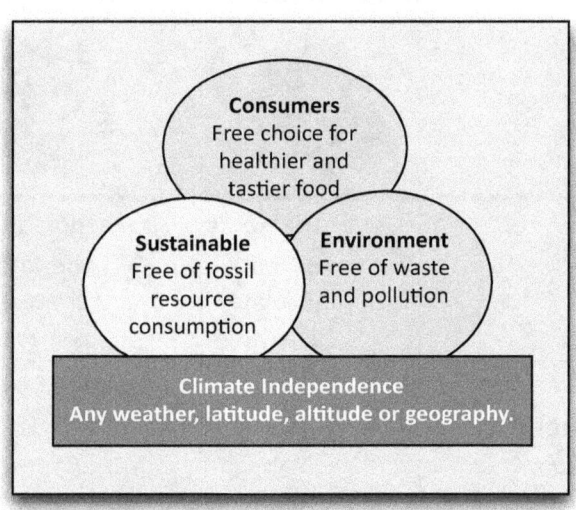

Scientific evidence and common sense show that foods low on the food chain consume significantly fewer resources. Consider the fresh water consumption of one pound of:

- Beef   26,400 gallons
- Corn   168 gallons
- Algae   0 gallons

A pound of algae provides more protein than beef and three times the protein of corn. Algae provide 50 to 500% more nutralence, essential nutrient availability and density. Micronutrients include vitamins, minerals, antioxidants and trace elements. Since algae cells are so tiny, they are immediately bioavailable to the body, which aids both digestion and nutrient assimilation. Bioavailability reduces the quantity of animal feed needed and reduces fecal waste.

Freedom foods are clean, healthy, and extremely low in fat and cholesterol. Consider the value proposition for chocolate cake.

Note: Freedom foods include the full spectrum of microorganisms such as algae, yeast, fungi, bacteria, archaea, protists, plankton and many others that grow in symbiotic communities.

Climate Independent Foods

Table 5.1. Industrial Food versus Freedom Food Chocolate Cake

Value	Industrial food Boxed Chocolate Cake Mix	Freedom food Chocolate Cake from algae compounds
High saturated fats	Yes	No, 80% less
High cholesterol	Yes	No, 90% less
Empty calories	Yes	No, 100% fewer
Pesticide residue	Yes	No, 100% less
Genetic engineering	Yes	No, 100% less
Preservatives	Yes	Minimal, 80% less
High protein	No	Yes, 100% higher
High macronutrients	No	Yes, 100% higher
High nutralence	No	Yes, 300% higher
Healthy omega-3s	No	Yes, 100% higher
Antioxidants	Few	Yes, 100% higher
Vitamins and minerals	Few	Yes, 100% higher
Trace elements	Few	Yes, 100% higher
Diminishes food cravings	No, increases	Yes, diminishes nosh
Consumes:		
Fertile soil	High	100% less
Fresh water	High	50 to 90% less
Fossil fuel	High	50 to 100% less
Fertilizer – mined	High	50 to 100% less
Pesticides	High	100% less

What are Freedom Foods?

Freedom to choose healthy food

Many modern consumers do not have access to healthy foods. The Centers for Disease Control published the Modified Retail Food Environment Index in April 2011 that shows that 9 out of 10 families lack access to retailers that sell healthy foods.[85] The Index reflects consumer access to retailers with fresh fruits and vegetables. Based on a range from zero (no food retailers that typically sell healthy food) to 100 (only food retailers that sell healthy food), the national average score was 10. Freedom foods can change the access problem by providing local foods that are fresh and healthy because these foods can be produced almost anywhere.

Modern consumers cannot currently make healthier food choices with freedom foods because these foods are not on the market yet. The USDA spends billions of dollars on fossil food research and subsidies, but largely ignores natural foods. Less than 1% of the USDA R&D budget supports organic production. The U.S. government has made small investments in algae research recently, but the focus has been biofuels, not sustainable and affordable food. NASA's $18.5 billion budget explores space but does not produce food.

Genetically engineered, (GE) foods may or may not bring health hazards. The USDA, FDA, and EPA have failed to require labels on GE foods. Lack of food labels combined with weak enforcement has enabled "GE creep." In slightly over a decade, GE foods have gone from zero to where today they make up a major portion of packaged foods. Over 90% of U.S. food grains such as corn and soy grow in GE monocultures that are refined into products loaded with fat, cholesterol, and calories devoid of essential nutrients, (empty calories). The USDA has approved 81 GE crops, while failing to deny a single proposal.[86] Applications pending propose to use transgenics to alter up to 30 genes simultaneously for a single crop.

Modern processed foods are high in fat and cholesterol. They cause obesity, diabetes, and a litany of Western diseases, including heart disease and cancers. Childhood obesity has increased 30% of the last 30 years and leads to diabetes. Diabetes is the leading cause of kidney failure, lower-limb amputations, and new cases of blindness in the

Climate Independent Foods

U.S. Diabetes is a major cause of heart disease and stroke and the seventh leading cause of death in the U.S.[87]

Freedom foods are healthy and can treat and in some cases prevent obesity, diabetes, and other diseases. These foods are naturally biodiverse, which eliminates the need for GE monocultures. Unlike fossil foods, freedom foods are clean, free of chemical fertilizer and pesticide, herbicide, and fungicide residue.

Natural resource efficiency

Our modern food supply is not sustainable because growing crops consumes massive amounts of fossil resources and then, too often, crops fail due to bad weather. Industrial agriculture wastes nearly half of applied resources even in good weather years. These resources percolate below the root zone or erode with wind and water and pollute fields and ecosystems.

Fertile soil serves the foundational natural resource for farmers and human societies. Industrial farming rips fields with cultivation and strips soils of nutrients and humus. According to the FAO, over the last 30 years, farmers were forced to abandon 33% of the cropland globally due to industrial agricultural practices.[88]

The fossil resources required to produce industrial foods will be gone in a few generations. Long before the fossil resources become extinct, they will become unavailable or unaffordable. Fossil resources including especially fertile soil, freshwater and fertilizers are already unaffordable to many farmers in India, China and Africa.

Organic food production is healthier for people and producers than industrial, but represents less than 3% U.S. of cropland.[89] Organic farmers avoid, to the degree possible, GE seeds, chemical fertilizers and agricultural chemicals and poisons. Unfortunately, organic farming methods cannot meet global food needs because organic production uses more fossil resources than industrial agriculture.

Organic production offers some resource advantages to industrial farming but gives no more weather tolerance than other forms of agriculture. Organic farmers often must use more cropland than industrial to produce the same amount of produce. Organics are

What are Freedom Foods?

equally sensitive to and consumptive of freshwater. Organic farmers do an excellent job at sustaining natural soil fertility but at the high time and energy cost of collecting, transporting, storing, applying and turning compost into the soil.

Freedom foods can use organic or industrial production methods. Ideally, growers use abundance methods that are organic and, in addition, use no or minimal fossil resources. Abundance growers may use brine water, which is a fossil resource, but plentiful supplies are available for many generations. Growers may use non-cropland and cultivate microcrops with renewable energy. Growers avoid the use of inorganic fertilizers by recovering sterilized waste stream nutrients.

Natural path

Freedom foods follow nature's path, exploiting the oldest and most efficient food production system on Earth. Algae and the diverse microorganism communities algae attract provide superior nutrition for people, animals, and plants. The tiny algae cells are packed with nutrients and are immediately bioavailable to consumers.

Corn produces its first gram of protein in about 120 days, a full growing season. Consumers and growers must wait another 365 days for the next harvest. Freedom foods growers produce the first gram of protein in about two weeks. Growers then harvest additional protein every few days, all year round. Each pound of algae flour delivers three times more protein than corn flour.

Microflora communities are similar to the organisms in our gut that break down food and aid digestion. Growers cultivate microcommunities in microfarms and train them to produce food, feed, fertilizer, and many other coproducts. The focus here is on algae, but a diversity of microflora may yield similar good foods with attributes superior to fossil foods.

These tiny biofactories run efficiently as they recycle and reuse the residual energy and nutrients from brine water or waste streams. Microfarmers use solar heaters and UV light to kill pests and pathogens before algae go to work to absorb wastewater organics.[90]

Climate Independent Foods

Higher nutralence

Succulence is the natural ability of succulent plants to absorb and hold water.[91] Algae demonstrate nutralence as the biomass concentrates nutrients at substantially higher levels than land plants.[92] The reach of their roots and the available soil nutrients limits the nutrient density of land plants. Algae avoid the root problem by living without roots.

Algae provide a low fat, low calorie, nearly cholesterol-free source of protein. Some algae, such as spirulina, contain up to 70% protein by dry weight – twice the protein of meat. Unlike meat, most algae varieties provide the full complement of nine essential amino acids. The low fat content, only 5-10%, is a fraction of other protein sources.

One tablespoon, 10 grams of algae delivers the same amount of:
- Calcium as 8 tbs milk, 32 tbs soybeans, 8 carrots, or 22 tomatoes.
- Magnesium as 40 tbs milk, 8 tbs soybeans, 9 carrots, or 6 tomatoes.
- Iron as 512 tbs milk, 8 tbs soybeans, 11 carrots, or 5 tomatoes.

Field studies show that algae nutralence, other vitamins and minerals are similarly 100 to 300% denser than field crop produce.[93]

Algae offer 22 times more calcium per tablespoon than tomatoes

The fibrous components of algae add bulk to the digestive tract reducing hunger pangs, transit time, and intestinal pathologies.[94] The total fiber content of algae (~6 g/100g) is greater than that of fruits and vegetables promoted today for fiber content: prunes (2.4 g), cabbage (2.9 g), apples (2.0 g), and brown rice (3.8 g).[95]

A chicken egg contains about 300 mg of cholesterol and 80 calories while providing the same protein as a tablespoon of the algae

What are Freedom Foods?

spirulina, which carries 1.3 mg of cholesterol and 36 calories. Algae are also an excellent plant source of glutamic acid, an amino acid that promotes intestinal health and immune function.

Each kilogram of algae biomass has two or three times the protein available from a kilogram of food grain. Algae concentrate many other nutrients at a multiple of the nutrients found in grains. Foods made from algae offer substantially more nutralence than food grains. Consumers benefit with more nutrients per calorie and per bite.

Per tablespoon, algae provide 10 times the beta-carotene available from the land plant that delivers the most beta-carotene, carrots. Beta-carotene are carotenoids that are highly pigmented (red, orange, yellow), compounds naturally present carrots and some other vegetables and fruits. Alpha, beta, and gamma carotene are considered provitamins because they can be converted to active vitamin A. Carotenes possess antioxidant properties and serve vital biological functions. Vitamin A deficiency leads to abnormal bone development, disorders of the reproductive system, xerophthalmia (dry eyes), night blindness and ultimately death.

Algae hold 10 times more vitamin B-12 and iron than beef liver. Vitamin B12 is an essential vitamin found in some fish, dairy products and animal liver. B-12 deficiency causes pernicious anemia, the inability to absorb vitamin B12 from the intestinal tract. B12 deficiency is common in the elderly, HIV-infected persons and vegetarians who are not getting sufficient B12 from their diet.

Natural biodiversity

Abundance growers have access to splendid natural diversity that enables them to grow biomass of 30 to 70% protein, depending on their target food or coproduct. Growers that want to maximize lipids (oils) may select a species that contains 40% lipids. Other growers may want to maximize production of carbohydrates, pigments, vitamins, minerals, antioxidants, cosmetics, medicines, vaccines or many other valuable coproducts.

Algae absorb a wealth of mineral elements that concentrate as about one third of its dry biomass. The macronutrients include sodium, calcium, magnesium, potassium, chlorine, sulfur and phosphorus while the micronutrients include iodine, iron, zinc, copper, selenium, molybdenum, fluoride, manganese, boron, nickel and cobalt.

Health benefits

Although very low in fat, algae offer an excellent source of the essential polyunsaturated fatty acids. The omega-6 and omega-3 fatty acids (ARA and EPA/DHA respectively) are necessary for normal metabolism, as they are the precursors to critical hormone-like, signaling molecules known as the eicosanoids. These short-lived messengers direct life-supporting functions such as blood clotting, inflammation, vasodilation, blood pressure and immune function. Only small amounts of ARA and EPA/DHA are needed daily (<1 g), and one tablespoon of algae can supply about half this amount.

Fish do not synthesize omega-3 in their oil. Algae synthesize the omega-3s that fish accumulate in their oil and that support human brain, eye and heart functions. Algae-based foods provide vital polyunsaturated fatty acids, omega-3, 6 and 7.

Medical research shows that omega-3 fatty acids reduce inflammation and may help lower risk of chronic diseases such as heart disease, cancer, and arthritis. Omega-3 fatty acids concentrate in the brain and are important for cognitive (brain memory and performance) and behavioral functions. Infants who do not get enough omega-3 fatty acids from their mothers during pregnancy are at risk for developing brain, vision and nerve problems. Insufficient omega-3 deficiency include fatigue, joint pain, poor memory, dry skin, heart problems, mood swings and poor circulation.

Algae contain a wide spectrum of prophylactic and therapeutic factors that include vitamins, minerals, amino acids and essential fatty acids. Algae provide the super anti-oxidants such as β-carotene, vitamins A, B, B-complex, C, D, E, and K, and a number of unexplored bioactive compounds.[96] These constituents stimulate numerous metabolic pathways and promote antioxidant, anti-bacterial, antiviral, anticancer, anti-inflammatory, anti-allergic, and anti-diabetic actions. Extensive medical research shows algae constituents promote vascular, mental, and intestinal health.[97]

What are Freedom Foods?

Phytic acid compromises the mineral availability from land plants, particularly legumes and grains, because the acid binds the minerals, rendering them unavailable for absorption into the blood stream. Phytic acid is typically absent in many algae species. Studies show that iron absorption is 3.5 fold greater for algae compared to rice.[98] Algae iron is easily absorbed by the human body because its blue pigment, phycocyanin, forms soluble complexes with iron and other minerals during digestion making iron more bioavailable. Hence, unlike iron derived from land plants, the bioavailability of algae iron is comparable to that of heme iron in meats.[99]

Algae nutrients, vitamins and minerals enhance physiological systems including the cardiovascular, respiratory and the nervous systems.[100] Algae components activate the cellular immune system including T-cells, macrophages, B-cells and anti-cancer natural killer cells.[101] Algae polysaccharides inhibit replication of several enveloped deadly viruses including herpes simplex, influenza, measles, mumps, human cytomegalovirus, SARS, and HIV-1.[102]

Algae's nutralence, antioxidants, enzymes and extracts, boost the immune system and enhance the body's ability to grow new blood cells. Algae are rich in phytonutrients and functional nutrients that activate digestive and immune systems. Algae compounds accelerate production of the humoral system (antibodies and cytokines), enabling the body to protect against invading germs.[103] Specific algae polysaccharides have demonstrated anti-atherosclerotic functions, reducing blood LDL cholesterol concentrations, and cardiovascular disease risk.

Research on humans and animals shows algae components have utility in the prevention and control of diabetes.[104] Other studies have demonstrated algae's therapeutic value for cholesterol management, blood pressure, heart disease and cancers.[105] Algae can moderate chronic inflammation that often precedes degenerative diseases. Algae provide therapeutic value for diabetes and fat metabolism.[106]

Research on mice shows algae delays the onset of motor symptoms and disease progression in ALS (Lou Gehrig's disease). Algae reduce inflammatory markers and motor neuron death.[107] Algae are calcium

Climate Independent Foods

rich and may protect against osteoporosis.[108] Recent research suggests algae activate human stem cells, which provide a spectrum of health benefits, including moderation of brain degeneration.[109]

New food supply

Freedom foods avoid the fatal errors with fossil foods.[110] Freedom foods can transform our food production system so that consumers could make healthy choices for themselves, producers, our planet and our atmosphere, Figure 5.2.

Figure 5.2 Freedom Foods benefit everyone

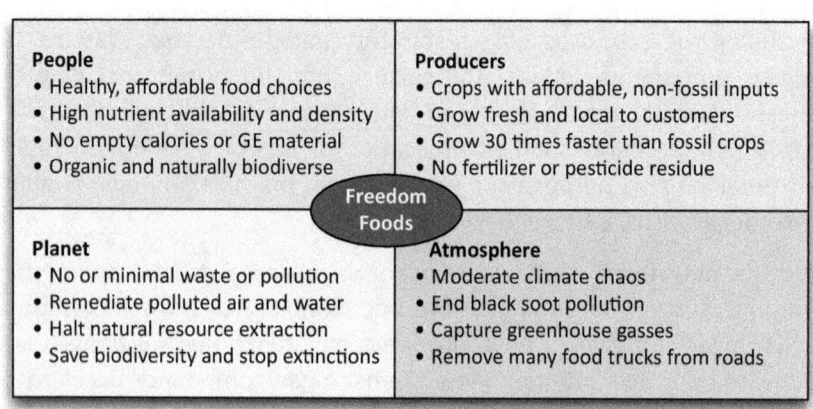

People	Producers
• Healthy, affordable food choices	• Crops with affordable, non-fossil inputs
• High nutrient availability and density	• Grow fresh and local to customers
• No empty calories or GE material	• Grow 30 times faster than fossil crops
• Organic and naturally biodiverse	• No fertilizer or pesticide residue

Planet	Atmosphere
• No or minimal waste or pollution	• Moderate climate chaos
• Remediate polluted air and water	• End black soot pollution
• Halt natural resource extraction	• Capture greenhouse gasses
• Save biodiversity and stop extinctions	• Remove many food trucks from roads

This new food supply will enable us to leave a positive legacy for our children – healthy, affordable food, clean ecosystems, breathable air, and abundant natural resources. Freedom foods offer a clean, naturally biodiverse and healthy alternative to fossil foods.

These foods can be grown locally or regionally and are superior in nutrition and taste, yet create minimal pollution or waste. The use of plentiful resources that will not run out presents the opportunity for food democracy, where everyone has access to good food or the affordable inputs to grow their own food. The use of plentiful resources means food supplies will be available and affordable for many generations.

What are Freedom Foods?

Freedom Foods vision:

> *Distribute the knowledge and capability for abundance methods globally to enable all people to grow good food and coproducts for their family and community locally.*

Distributed, local food production produces fresh foods that do not need preservatives and avoid the high dollar and energy cost of transportation.

Algae already in our food supply

Most consumers are surprised that food processors already integrate algae components in our food supply. A market basket test at Arizona State University examined the non-fresh produce items in 10 typical shopping carts for foods containing algae components. About 72% of the foods and 88% of the cosmetics contained algae constituents.

Algae Flour makes delicious Freedom Foods

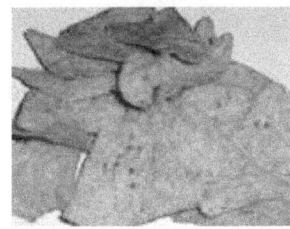  

Beer and soft drinks use algae as a clarifier. Algae cell walls contain carrageenan used as a stabilizer or emulsifier found in dairy, confectionary and bakery products. Alginates provide alginic acid from brown algae, which thicken liquids and make them creamier and more stable over wide variations in temperature, acidity, and time.

About half of the alginate harvested from kelp goes into ice cream and other dairy products to make them smoother and prevent ice crystals. Alginates also keep toothpaste and lipstick from going dry.

Algae contain considerable agar, a polysaccharide that solidifies almost any liquid. Agar acts as a colloidal agent used for thickening, suspending, and stabilizing soups, stews and canned goods. Agar is used as a substitute for gelatin, as an anti-drying agent in breads and pastries and also for gelling and thickening. Agar enhances processed cheese, mayonnaise, puddings, creams, jellies, and dairy products.

Algae makes delicious Freedom Foods

  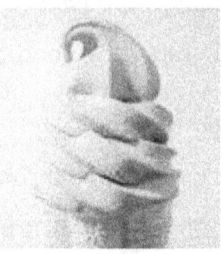

Algae naturally form flours that can substitute for food grains such as rice, corn, wheat or barley. Algae flour makes chips, dips, breads, tortillas, crepes, cakes and pretzels that have superior nutrient profiles to their cousin foods with substantially lower fat and cholesterol. Of course, people will not choose the algae-based food models unless they taste similar or better and are affordable.

Algae milk can substitute for dairy, soy, or almond milk while providing higher protein. Many people are lactose intolerant and others are bothered by soy and nut allergies. Algae milk induces neither intolerance nor food allergies. Algae sugars can substitute for cane, beet or corn sugars. Solazyme recently announced new algae chocolate improved mouth feel, yet 85% fewer calories.

Texturized algae made into vegie burgers pack the savory umami taste. Currently, the extremely popular Umami hamburger chain in Los Angeles must add MSG to their beef to gain the attractive hearty flavor. When the Umami firm uses algae, they will be able to provide great flavor with natural savory taste from algae.

Consumers in a few years will have a choice for hamburger. A beef hamburger, with supply costs and environmental takes, may cost $35. A cultured meat burger, where the meat is cultured in a laboratory without and animal, may cost $25. (Currently, cultured meat has the consistency of snot.) The algae burger alternative will cost about $8 and offer all the health benefits associated with other freedom foods.

What are Freedom Foods?

Consumption of the algae burger will allow the eater to leave a net zero ecological footprint, similar to a butterfly.

Availability

Sea vegetables and other algae products are popular for their nutrition, color, taste and texture in Asian Pacific Rim countries.

Sea Vegetables

Many high-end European restaurants offer algae appetizers, entrees, main courses and deserts. In the U.S., freedom foods are available currently only in Asian markets and health food stores, where they sell as food supplements. Whole Foods, Trader Vic's and Costco recently began selling algae snack foods. Sushi bars combine several forms of algae with other seafood and rice.

Roasted Seaweed at Costco

Food companies have investigated algae-based foods for decades but they faced a pair of showstoppers with freedom foods – supply and demand. Growers have no incentive to grow algae when terrestrial foods are so heavily subsidized and cheap. Consequently, food-processing companies have no suppliers.

Climate Independent Foods

General Foods, Nabisco, Frito-Lay, Borden, Dial or Quaker Oats cannot manufacture an algae product today because there is no market. Consumers do not select these foods because they know little about the health and environmental value of freedom foods.

Algae in restaurants

Why does the number one rated restaurant in the world, NOMA, serve algae? Rene Redzepi chef, forager, and owner of NOMA, in Copenhagen, made the March international cover of *Time Magazine* for his excellent innovations in locally sourced and foraged cuisines. His extraordinary NOMA cookbook shows beautiful pictures of algae garnishing and supporting his world-class servings.

Rene Redzepi and Umami Burger

The human tongue has five taste buds: sweet, sour, bitter, salty and umami. Umami, the savory or hearty taste comes from taste buds in the middle of the tongue that give a savory or hearty taste – not to just the food eaten directly, but also to the accompanying food. Many upscale restaurants offer algae directly or accompanying hors d'oeuvres and main courses to add the desirable savory taste. One of the most popular new restaurant chains in Los Angeles, Umami Burger, has found great success promoting the umami taste.

Dozens of excellent algae recipes and algae foods are available at www.AlgaeCompetition.com and on Epicurious.com

Algae flour

Solazyme in partnership with Roquette Nutritionals announced in April 2012 the availability of their Algalin™ flour. Algalin flour looks and acts like flour, but is actually a lipid substitute that is similar to

What are Freedom Foods?

olive oil. Product recommendations explain that the flour can substitute for eggs, cream, milk, vegetable oils or other lipid sources.

Algalin flour provides improves nutritional profiles in many foods such as bakery, beverages and frozen desserts. Algalin flour acts as a whole food ingredient and delivers very low saturated fat, no trans-fats and no cholesterol. It reduces calories considerably, as well as provides fiber and protein, while providing the same overall mouthfeel and consistency as a full fat food.

Algalin™ brownies, flour, oil and nutritional drinks

Altein™ Algalin protein provides over twice the levels per gram as corn or soy protein. The product has about 20% dietary fibers, 10% lipids and a wide array of trace minerals and micronutrients. The vegan product is marketed as gluten-free, non-allergenic, sustainably produced with highly digestible protein.

Since the Solazyme algae products are made with fermentation and require sugar for the energy instead of photosynthesis, some may argue the sustainability claim. However, Solazyme works very hard on their sustainability and has invested in farms to grow plants that can supply the necessary sugars.

New demand for substitutes

One of the most interesting recent food stories belongs to Girl Scouts and palm oil. Two Girl Scouts created an initiative to end the use of palm oil in Girl Scout cookies because palm oil farming causes rainforest deforestation endangers thousands of animal species and contributes to human rights abuses. Over 70,000 people have signed their petition to stop the use of palm oil in Girl Scout cookies. The two

Climate Independent Foods

girls have been featured on numerous news and talk shows, and they were recently honored with the United Nations Forest Heroes Award for their work in saving rainforests.

Two Girl Scouts work to end Palm Oil use in Scout Cookies

Fish oil may be right behind palm oil. Several environmental and scientific groups are working to end the unsustainable catch of small fish for oil animal feed and fish oil, omega-3 fatty acids.

"Little Fish, Big Impact," financed by the Lenfest Foundation through the Pew Charitable Trusts, reports how small fish catch has increased to 31 million metrics tons or 37% by weight of all fish harvested globally.[111] Thirteen eminent scientists invested 3 years reviewing the scientific literature and analyzing computer models of food webs in 72 marine ecosystems around the world. They found that 75% of the ecosystems they studied had one or more large predators for which forage fish made up at least half their diet.

Forage fish are algae feeders that deliver essential nutrients absorbed and synthesized by algae to their consumers higher in the food web. Forage fish are a critical link in the food chain and are consumed by just practically every fish, bird and mammal in the ocean, including tuna, dolphins, puffins and penguins.

The task force at the Institute for Ocean Conservation Science at Stony Brook University led by Ellen Pikitch estimated that forage fish are worth more than $11 billion to the oceans. Left in the ocean, forage fish are worth twice as much as when processed for fish oil and aquaculture.[112]

What are Freedom Foods?

As companies look for sustainably sourced oils, algae oils will become popular rapidly because the value proposition has already been communicated to consumers, save the dolphins and puffins.

Save dolphins and puffins, eat algae omega-3 oil

Commodity prices

Constantly rising commodity prices are prompting demand for cheaper and more sustainable substitutes for terrestrial crop commodities. Commodities that have been relatively constant for years are rising.

Higher commodity prices translate quickly into higher food prices. The Arab Spring ignited over food prices and food availability in Tunisia, Libya, Yemen and Egypt. Food prices exceed 40% of family income. Global food demand is now so high and rising so fast and reserves are so low that price sensitivity to crop yield drops or failure has become extreme. As food prices continue to rise, more countries will face uprisings and possibly regime changes.

Global climate chaos devastates crops. China, India and Pakistan lost millions of acres of farmland in the severe 2011 and 2012 floods. Communities also lost the critical infrastructure for crop production, roads, bridges and irrigation systems.

Algae can play a major role in sustainable cultivation of substitutes for food grains, vegetable oils, fish oils, and other sources of protein, including meats. Business opportunities abound in the algae realm of sustainable food and energy products.

Freedom foods adoption

Skeptics predict that consumers will not accept freedom foods, despite their many advantages. The science of consumer behavior offers two strategies to change people's food consumption:

1. **Push** people to eat lower on the food chain with influence strategies, such as carefully explaining the health and environmental benefits.
2. **Pull** occurs as consumers demand and buy nutritious and delicious foods that appear similar, yet taste better than their equivalent fossil food.

The push strategy called cognitive conditioning, works only for a tiny percentage of consumers. Most people try diets or practice good food habits for a brief period, and then become recidivists. They revert to their favorite comfort foods. The statistics on diets confirm that most do not work.[113] Food behaviors are central to family, culture and society, and are extremely difficult to change.

Push strategies have failed scientists and practitioners repeatedly. Many fat people are experts on nutrition. Repeated studies show that knowledge alone fails to change behavior. People change their behavior when they find a preferable substitute.

Push strategies cannot work because fossil foods monopolize our fields, stores, refrigerators and plates today. The only choice most consumers can make currently is industrial versus organic foods or meat versus vegetables. Industrial foods control 97% of the U.S. market and 94% in Europe. Unlike the U.S., the European Union offers organic growers subsidies for the social, environmental and health benefits provided by organic foods. Freedom foods offer a fossil free alternative, aligned with organic methods, and will give consumers another choice – once they are available.

The Freedom Foods Revolution proposes a pull strategy where food low on the food chain goes into products that appear similar to traditional foods but with ingredients low on the foods chain that are natural, healthier and provide superior sensory pleasures. To gain widespread consumer adoption, freedom foods must offer

What are Freedom Foods?

more than health and environmental benefits. Consumers must perceive them as offering a bundle of positive attributes. Each additional positive characteristic will accelerate consumer adoption.

When consumers have the freedom to choose foods with superior sensory appeal and better nutrition at a reasonable cost, they are likely to choose freedom foods.

Freedom foods will attract some consumers for the health benefits from eating low on the food chain. Others will choose freedom foods to avoid GE crops, preservatives and chemical residues. Some will want freedom foods to lighten their ecological footprint. The majority will probably make their choice based on superior nutrition and taste.

The Freedom Foods Revolution plans to spark demand by educating consumers, which will motivate new suppliers. The new freedom foods industry will offer thousands of engaging entrepreneurial opportunities for growers, suppliers, restaurateurs and chefs. Please see our global social collaboratory, AlgaeFuture.org.

The next chapter explores the amazing story of algae in human food history.

Chapter 6. Algae in Human Food History

Did algae make us human?

Many people assume algae lives up to its role expressed by uninformed journalists who repeatedly use terms such as slick, slime, scum, and other terms with negative attributions. Ironically, just the opposite is true. Algae's attractive taste may have helped us become human by attracting our ancestors to algae and the Omega-3s that sparked brain enlargement.[114]

Algae played pivotal roles in human evolution and survival. Early human societies evolved along coastlines, rivers and lakes and depended on algae for food and medicines.[115] The high nutralence biomass was plentiful year-round and easy to harvest. Many groups ate algae directly and probably ingested algae in their drinking water.[116]

Algae provided a rich and nearly complete source of nutrition – a complex blend of nutrients that no other food source, plant or animal, could offer.[117] Algae were analogous to a modern-day vitamin supplement – but actually, algae are a more robust, natural, and inclusive blend of healthful nutrition.[118]

Algae are a superior protein source, particularly the red, green and blue-green algae, which are as high as 70% protein (dry weight), which is higher than and corn, 23%.[119] Algae protein content is highest in the late winter and early spring, which is advantageous when terrestrial plant food sources are scarce. Algae nutralence benefited our ancestors year round by preventing many of the nutrient deficiencies that plague modern human societies.

Algae in Human Food History

Did Algae make us human?

Possibly the most interesting unanswered question in science is: "How did we become human while our contemporary Homo cousins became extinct?"

Our ancestor's attraction for the sweet taste of algae may have played a significant role in becoming human. Strategically placed evidence on our tongue provides fascinating clues that science has so far missed. Our pre-human ancestors made a significant, possibly accidental, decision to ingest algae, which may have led to the evolution our large brain and enabled *Homo sapiens* to evolve, thrive and rise to the top of the food chain.

Scientists agree that human brain enlargement—encephalation—differentiates *Homo sapiens* from our ancestors. Our pre-human ancestors evolved from chimpanzees around 8 million years ago (mya) but very little happened to the brain for the first 6 million years. About 2 mya, brain enlargement began and by 1.5 mya, the humanoid brain was three times the size of chimpanzees. What happened to our ancestors during this half a million years of evolution? Humanoids brains grew larger a million years before cooking fires or hunting weapons were invented.

Larger brains require substantially more energy because brain mass consumes 16 times more energy than muscle mass. Therefore, our ancestors traded muscle for brains. Something triggered brain enlargement and the logical answer was a change in diet. Current theory posits that our ancestors moved from the primate diet of leaves, bark, insects and occasionally fruit to a more diverse diet, higher on the food chain that included game meat.

However, moving up the food chain to hunt game meat would have been problematic for slow, scrawny and pitifully weak hominids that the fossil record shows had Lucy's stature of only 3.5 feet. Relative to predators, they had poor senses; including sight, hearing and smell. Had they decided to hunt, they would have quickly entered the food chain rather than dominating other animals.

Climate Independent Foods

Other scientists suggest early hominids practiced scavenging to gather game meat. They may possibly have found bones that could be pounded for the marrow. Since both predators and scavengers 2 mya were twice the size they are today, hominid scavengers would have been very lucky to find a carcass. The hominid scavenger probably would have run out of luck hauling the food back to camp because stealth predators were numerous and ferocious. Early hominids probably were subject to annual predation at the same rates at which living primates living under natural conditions are today—roughly 8% of their local population.

Independent of the physical improbability, the scavenging scenario is unlikely due to our weak stomachs. Meat begins putrefying immediately after death. Spoiled meat attracts parasites, insects, maggots, worms, bacteria and other microorganisms that would have been just as fatal for early hominids as they are for humans today.

Anthropological evidence shows humans evolved in East Africa along the Rift Valley Lakes that are unique ecosystems. These ancient, high alkaline soda lakes have large natural stands of spirulina. Spirulina flourishes as a nearly pure culture in the high pH alkaline water. The blue-green algae do not fix nitrogen and the biomass is safe to eat.

Moving down the food chain

The dance theory suggests that we became human with a waltz. Our ancestors probably took two small steps down the food chain before taking a big step up to hunt game.

Early hominids' first step was probably down the food chain where they ingested algae in their drinking water. Terrestrial plant foods available to early Homo were largely hard, dry and bitter. Wild seeds, nuts, roots and leaves would have provided little variation in terms of the four classic tastes; sweet, bitter, salty or sour. Sweet would have been largely missing from their diet, except for the rare occurrence of fruit. Salty would have been missing except for salt licks and they provide an unpleasant intense salty taste, independent of food. The sweet, green algae water would have provided a subtle sweetness and occasionally saltiness. Compared to the rest of the diet available to early Homo, sweet green water would have been very attractive.

Of course, early hominids did not intentionally ingest algae to enlarge their brains. Evolution is not volitional and our ancestors probably could not even see the algae in the water because the cells were too small. Algae first become visible as a light cloud. As the cells proliferate, they turn the water green but the individual cells are not visible to the naked eye.

The sweet taste in algae probably attracted *Homo*. Our ancestors drank from the top of the water column, drinking directly with their mouth or by using their hand as a scoop. Algae grow and concentrate on the top of the water column, which made drinking an effective method of harvesting algae.

Algae in drinking water would have acted as a natural food supplement to supply the essential nutrients, vitamins and antioxidants, especially the omega-3 fatty acids that provided the green spark for encephalation.[120] Larger brains differentiated our ancestors from their cousins and enabled higher cognitive skills that aided survival.

On the Rift Valley Lakes where early Homo developed larger brains, algae grow to such concentrations that the thick green biomass forms mats on the top of the water column. The wind aggregates algae naturally on the lee side of lakes. These concentrated algae mats would have been easy to gather for supper or store for later use.

Algae Mats

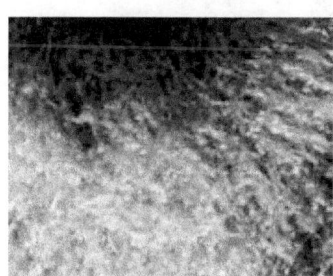 

Spirulina is the bestselling algae nutritive supplement on the market today because it provides a complete set of essential nutrients, including vitamins, minerals, trace elements and antioxidants.

Climate Independent Foods

As brains became larger, hominids probably took another small step and exploited the lacustrine ecosystem. The edges of rivers and lakes are filled with nutralent rich aquatic plants, fish, crustaceans, amphibians and birds. Many of these creatures are what they eat and simply concentrate the nutrients from their predominately algae diet. Fish and other aquatic creatures do not synthesize omega-3 fatty acids, they concentrate omega 3 from its source – algae.

Harvesting algae and the meat of algae feeders from rivers and lakes would have allowed early Homo to grow a larger brain and body. Once hominids had sufficient strength and stamina, they would have been ready to take a big step up the food chain to the savanna where they could hunt meat with their strong brains, eyes and bodies.

Taste bud evidence

The human tongue has specialized taste bud receptors for a fifth taste – umami. Umami is a rich flavor constituent found in some protein-rich foods. The unique umami taste which means "good flavor" in Japanese has been isolated and marketed as monosodium glutamate (MSG). Umami translates to brothy, meaty or savory. The unique taste is induced by three proteinogenic amino acids: glutamic, inosinic and guanylic. The savory taste would have been largely absent from early hominid diets and would have been very attractive.

Umami

Glutamate plays a key molecule in human cellular metabolism. Proteins are broken down by digestion into amino acids, which serve as metabolic fuel for other functional roles in the body. Glutamate is the abundant excitatory neurotransmitter in the vertebrate nervous system and regulates several brain functions.

Glutamate's role in body functions is so critical that some animals, including humans, evolved a special taste bud call the mGluR4 receptor to taste glutamate. This umami taste is most concentrated in high protein algae and algae feeders such as fin and shellfish, which would have made these foods quite attractive to early Homo. (Today the umami taste can also be found in milk, aged cheese and some meat products.) The combination of sweet and umami tastes that were not available to our ancestors from terrestrial foods would have provided a strong incentive for early Homo to eat algae.

Color plays a sizable role in motivating appetite. Most seeds, grains and roots are a dull brown and neither pleasing nor appetizing. Algae have colorful pigments the plant uses to collect solar energy and drive photosynthesis. These pigments provide a spectrum of natural colorings to foods. Unlike modern synthetic colorings, algae pigments are not only colorful but also nutritious. Sea vegetables or local freshwater algae would have added attractive color when consumed on their own or when blended with other foods.

Among the many edible algae that would have been plentiful for early hominids in and around Africa were arame, alaria, seawhip kelp, chlorella, dulse, hijiki, karengo, kombu, nostoc, nori, ogo, sea lettuce, sea palm, spirulina and wakame. Algae components would have been available also including agar, alginates and carrageen as well as carotenoids (pigments) such as luten, phycocyanin, zeaxanthin, astaxanthin and phycobiliproteins.

Algae were probably our first food because algae were the best tasting, best texture and most colorful food available. Algae were the most plentiful when humans began gathering food. Algae delivered the highest nutralence of any available food source. Most edible algae provide significant protein and a rich assortment of macronutrients as well as essential vitamins and minerals; especially phosphorus, potassium, iron, selenium, copper and zinc.

Climate Independent Foods

Human migration

As early humans migrated out of Africa 70,000 years ago, they followed coastlines where macroalgae – seaweeds and sea vegetables – were plentiful. Anthropological records show caves used during human migration hold traces of sea vegetables and algae cuds. Algae cuds come from seaweeds that are chewed like gum to release the rich nutrients bound in algae cell walls.

At low tide, hominids could harvest algae easily and dry it quickly in the sun. The light produce probably served as the first wampum in trade because it was easily transportable. Algae wampum offered a side benefit; a hungry trader could eat the product.

Algae manage insulin release in the blood stream and create a feeling of satiety or fullness that would have been extremely valuable for hominids. Mothers had to carry their children and babies crying from hunger pangs would have created a predator threat to the mother and the tribe. Feeding infants a little algae would have helped brain, heart and eye development from omega-3 fatty acids as well as given them a comfortable feeling of fullness so they did not cry.

Convenience food?

Algae probably provided our ancestors with the original convenience food. Algae can be eaten fresh or dried and stored for up to two years. Dried algae turns dark blue, green or black. When reconstituted in water, algae recover the similar bright red, yellow, green or purple color it had when it was fresh.

Terrestrial foods were dry, hard, bitter and starchy. Land plants were difficult and risky to gather due to stealth predators. Algae offered a fresh, soft, delicious taste and were easily accessible. In many locations, algae were harvestable year round, which would have been a tremendous advantage when terrestrial crops were dormant or not producing.

Early humans probably rubbed algae oil on their skin for sun protection. Algae add moisture and speeds the recovery from wounds, burns and bruises. Algae's high antioxidant activity protects skin from inflammatory reactions and sun damage.[121] Pacific Rim

Algae in Human Food History

societies have been using algae for natural foods and remedies for centuries because they are effective. Organic chemists, medicinal chemists, biologists, and pharmacists are currently developing new anti-inflammatory and anti-cancer medicines from algae.[122]

The first written mention of algae was in Korea, about 57 BC. The Samgukyoosa contains passages that record gim, (today called nori) as part of the dowry for Shilla royalty. Members of the Chinese Court, around 1,100, harvested and reserved a specific algae variety for the Chinese Emperor. The Japanese reserved another algae variety for the Samurai, the Japanese nation's fiercest warriors.

The Aztecs used algae for food, medicine, trade and religious ceremonies. Indigenous people along coastlines or lakes have harvested natural stands of algae for millennia for use as food, feed, medicines and trade. Algae probably protected our ancestors against many diseases that plague developing countries today. Algae protects against scurvy, xerophthalmia (blindness from vitamin A deficiency), goiter, arthritis, diabetes, mental retardation and others. The Chinese have used algae for medical purposes for centuries because these natural remedies are safe and effective.

Roman farmers and soldiers valued seaweed as an animal feed supplement. Roman military officers fed algae to their horses to improve color and sheen to their coats as well as health and stamina. Farmers and gardeners used seaweed as a soil amendment to improve the color, taste and texture of produce. The classic red color of Roman military tunics came from pigments extracted from an algae-lichen crust known as urchilles. Wealthy Roman women used the pigments on their cheeks as rouge.

Chinese Olympic athletes have consumed algae for decades because, like the Samurai, algae nutrients enable them to train harder and longer. The therapeutic elements provided by algae allow the athletes to recover from injuries faster. The Chinese have a secret algae cultivation center near their Olympic training center that enables the athletes to eat the algae fresh. Fresh algae maximize the availability of micronutrients, especially antioxidants.

Climate Independent Foods

U.S.A Olympian Lee Evans won four gold medals at the 1968 games in Mexico and held of four world records in track and field. Evans ate spirulina because the algae improved his training and improved his speed and endurance. Geronimos Dimitrelos was competing for a place on the U.S. Olympics team in Greco Roman wrestling when he broke his wrist severely. He was so impressed with how algae speeded his recovery, that he started a new algae company, Algae2Omega to grow algae nutritional products.

NASA recommends algae foods for space flights because research shows algae improve mental acuity, eye sight, digestion and stamina. Algae build the immune system and moderates inflammation. Algae provide another benefit for space flight, habitat renewal. Algae can absorb all the exhaust gasses produced inside the living area and return pure oxygen to the air. Algae can pull all the organic wastes from the liquid and solid waste streams and repurpose them in food, feed, fertilizer or medicines.[123]

Spirulina drink

Spiralps® (spiralps.ch) is a new type of natural soft drink containing fresh spirulina, organic fruits and alpine herbs. The fresh *spirulina* in Spiralps, introduced in Europe in 2012, tastes great, better than dried algae and is nutritionally superior. Processing uses high-pressure pasteurization system that avoids heating the spirulina.

Spiralps® and Controlled Environment Growing System

The next section examines the nature of this exceptional plant.

Chapter 7. What makes Algae Special?

Algae, nature's oldest and possibly best food.

The foundation for freedom foods and abundance growing methods relies on a plant that used 3.7 billion years of evolution wisely to develop strategies to:
- Adapt to acute searing and freezing temperature spikes.
- Survive the extremely hot temperatures of early Earth.
- Grow in ocean, brine, saline, waste and fresh water.
- Live through brutal electrical, wind and ice storms.
- Use either organic material or sunshine for energy
- Go dormant when conditions degrade, yet survive.
- Grow rapidly at any latitude, longitude or altitude.
- Grow faster than any other plant on Earth.
- Thrive using no extracted fossil resources.
- Maximize productivity per unit of space.

All living organisms evolved from algae so it should be no surprise that algae contain all the essential nutrients for life and vitality. A few grams of algae a day act as a nutritional supplement, providing the essential nutrients, vitamins, minerals and antioxidants.

Climate Independent Foods

Microalgae

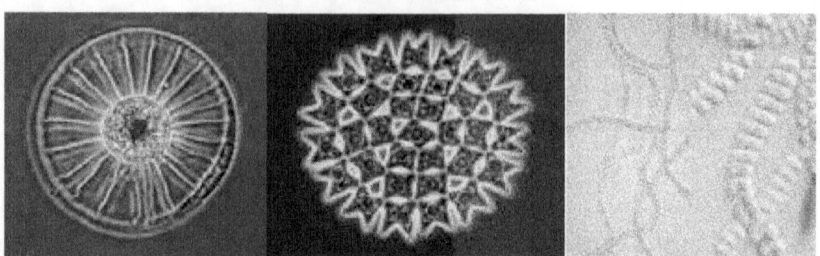

The first algae (singular) cell was among the earliest life forms on Earth, probably about 3.7 billion years ago in oceanic environment synthesized by abiotic, high-energy processes including lightning, ultraviolet radiation and pressure shock. The atmosphere was anaerobic with high levels of methane, hydrogen and ammonia but no oxygen.

Algae, often called microscopic phytoplankton, grow in most bodies of water, moist places, on and in trees, and even in rocks. This little plant provides the foundation for the food chain, feeding both microbial and animal plankton; zooplankton and fish. Subtract algae and phytoplankton from the water column and fish, shellfish, reptiles and other aquatic creatures cannot survive.

Algae grow all over the Earth in marine or fresh water habitats or on land, when moisture is available. Unlike land plants that die without water, algae simply go dormant, wait for moisture, and then begin their rapid development. Algae grow to the limit of the nutrient supply, and then pause or go dormant until conditions improve.

Since algae form the bottom of the food chain, everything around acts as a predator. Algae's strategy to predation is brilliant – grow faster than consumers can eat. A single alga cell may produce one million offspring in a day. At night, algae take a well-deserved rest in a phase called respiration. While an individual alga cell is not visible, algae communities appear first as a cloud and then as tiny specks that are cell clusters. Some algae aggregate to form structures, such as filaments, globes, wheels or with spirulina, spirals.

Algae break the rules for plant classification because they evolved in many different forms – cells, multicellular plants, bacteria and in

What makes Algae Special?

nearly infinite combinations. While the various species share certain characteristics, different algae display extraordinary variety in shape, size, structure, composition and color.

Many species are single-celled and microscopic including phytoplankton and other microalgae while others are multicellular and may grow large such as kelp and Sargassum. Phycology, the study of algae, includes the study of prokaryotic forms known as blue-green algae or cyanobacteria. Some algae also live in symbiosis with lichens, corals and sponges. The basic single-celled organism, algae, has the general appearance illustrated in Figure 7.1. The University of Montreal, U.C. Berkeley, University of Texas and others host culture collections of algae species with descriptive details and pictures.[124]

Figure 7.1 Algae Cell

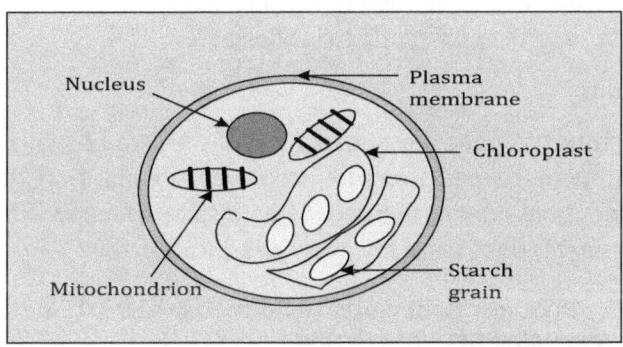

Eukaryotic green algae (Greek for "true nut") plants have cells with their genetic material organized in organelles. They create discrete structures with specific functions and have a double membrane-bound nucleus or nuclei. The prokaryotic cells of blue-green algae, cyanobacteria, contain no nucleus or other membrane-bound organelles.[125] Algae can be lively little critters even though they are not animals. Many can swim, such as dinoflagellates that have little whip-like structures called flagella. Some use the flagella to pull or push themselves through the water. Some algae squish part of their body forwards and crawl along solid surfaces.

Other species are made of fine filaments with cells joined from end to end. Some clump together to form colonies while others float independently. Seaweeds may grow in nearly any shape such as cones, tubes, filaments, circles or they may imitate the shape of land plants. Seaweeds developed in parallel evolution with land plants.

Algae Cell Walls

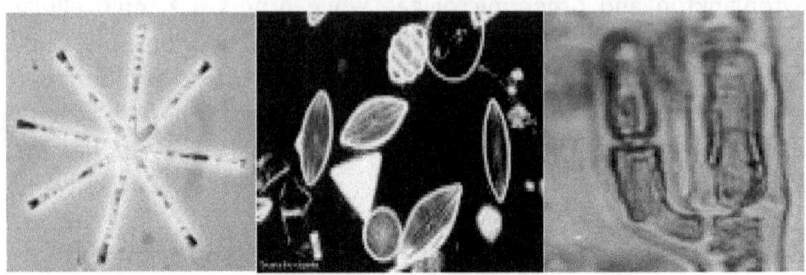

Major steps in cell complexity occurred with the evolutionary progression from a virus to bacterium and then from the prokaryotic cells of bacteria to the eukaryotic cells of algae.

Classifications

The major groups of algae have been distinguished traditionally based on pigmentation, shape, structure, cell wall composition, flagella characteristics, and products stored. Algae display so many variations they express exceptions to nearly every classification rule.

Diatoms, stoneworts and dinoflagellates

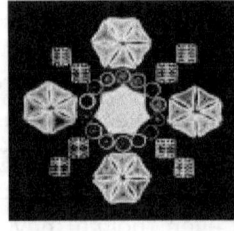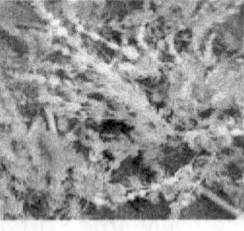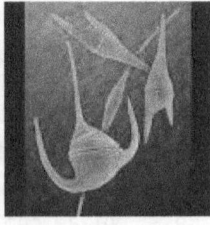

Prokaryotes, known as blue-green algae or cyanobacteria, evolved first, 3.6 billion years ago, creating oxygen in our atmosphere so higher life could evolve. After a billion years building up oxygen, and in a more oxygen rich environment, eukaryotic algae with nucleus evolved. The addition of a nucleus and cell wall were huge evolutionary advances. Cell walls enable algae to protect itself from the surrounding environment, typically water and pressure, called

What makes Algae Special?

osmotic pressure. Cell walls regulate osmotic pressure produced by water trying to flow in or out of the cell through its semi-permeable membranes due to a differential in the solution concentrations. Algae typically possess cell walls constructed of cellulose, glycoproteins and polysaccharides. Some have a cell wall composed of silicic (silicon) or alginic acid.

The broad algae classification includes:

- Bacillariophyta – diatoms
- Charophyta – stoneworts
- Chlorophyta – green algae
- Chrysophyta – golden algae
- Cyanobacteria – blue-green
- Dinophyta – dinoflagellates
- Phaeophyta – brown algae
- Rhodophyta – red algae

Each species has its unique constituencies and produces various levels of each compound during its life cycle. Producers may stress algae by withholding a specific nutrient or changing the temperature or pH to prompt algae to overproduce the target compound such as lipids.

Red algae are a large group of about 10,000 species of mostly multicellular, marine algae, including seaweed. These include coralline algae, which live symbiotically with corals, secrete calcium carbonate and play a major role in building coral reefs. Red algae such as dulse *(Palmaria palmata)* and laver (nori/gim) are a traditional part of European and Asian cuisine. Red algae are also used to make many other products such as agar, carrageenans and other food additives.

Productivity

Scientists have known algae's food value for centuries and food potential for at least 100 years. Consider the annual protein production per acre for food grains calculated using half its theoretical photosynthetic capacity, Figure 7.2. Algae provide a superior set of vitamins and minerals than found in land plants. Algae are not a full solution for malnutrition because the biomass is freedom on calories. Fortunately, calories are cheap and easy to add to a diet.

Algae flours are extremely malleable in the sense that algae can substitute for wheat, corn, rice or soy products while providing higher protein and a higher quality nutrient profile. Algae foods may include protein-rich milk, ice cream, chocolate (with superb taste and 80% less fat), baked goods of any size, shape or texture such as tortillas, crackers or cakes. The biomass may provide texturized vegetable protein with added fiber or extruded to make additives for meats that improve moisture retention and increase protein while lowering fat and cholesterol.

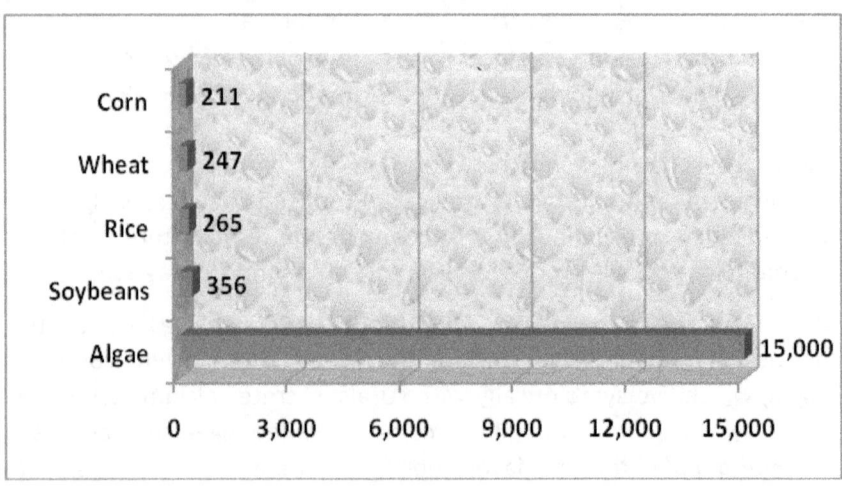

Figure 7.2 Algae Protein Production Potential – Pounds per Acre per Year

Processing algae can match the form of nearly any food such as pasta, pesto or protein bars. Decades of food processing experience with terrestrial crops that have an unappealing natural taste, such as soybeans, make it easy to add flavors, textures (fibers) and aromas.

Land plants have specialized cells for moving nutrients and for reproduction that algae do not need. Algae are distinguished from the higher plants by a lack of true roots, stems, or leaves. Some seaweed appear to have leaves or trunk but they are pseudo leaves made up of the same cellular structure as the rest of the plant.

Algae use nitrogen to manufacture amino acids, nucleic acids, chlorophyll and other nitrogen compounds. Many cyanobacteria species are able to fix nitrogen absorbed from the air, as well as from water, in a process known as diazotrophy. Since the atmosphere is

What makes Algae Special?

78% nitrogen, nitrogen fixing is a strong competitive advantage for growth because water-based nitrogen is often limited.

Nitrogen fixing also means that the plant biomass has value as a low energy input, high nitrogen fertilizer because algae fixes nitrogen naturally, without added energy. About 90% of the cost of commercial synthetic fertilizers comes from the energy, typically natural gas, used to extract nitrogen from the air.

Variation

Algae range from microscopic single-celled organisms to multicelled organisms and to 60-meter kelps. These plants thrive all over the world in marine and fresh water environments − nearly any moist environment. Terrestrial algae grows in all types of soils where they can capture nitrogen from the air that can be used through the roots of plants. They may be free-living or live in symbiotic association with a variety of other organisms such as lichens and corals.

Algae Shapes

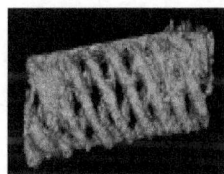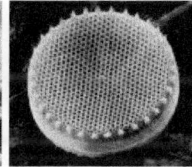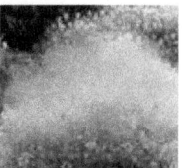

Each species may exhibit multiple strains with unique characteristics. A single strain may display completely different structural expression and composition in different growing conditions with variations in light, temperature, nutrients, mixing or water pH.[126]

Macroalgae

About 10% of algae species are macroalgae, (seaweed and sea vegetables) such as kelp that can grow to 60 meters. Most are microalgae that occur in every color, shape and small size imaginable.

Away from the oceans, most algae grow in, on or among the roots, trunk and leaves of land plants. A handful of local dirt may hold several billion alga cells and over 100 algae species. Land plants need algae to break down chemical fertilizers so they are bioavailable and

Climate Independent Foods

absorbable by the plant roots. Land plants and algae work symbiotically as algae supplies nutrients and plants provide a protected area in which to grow. Algae also support symbiotic relationships with mosses, fungi, yeasts, lichen, corals and sponges.

The next chapter introduces algae cultivation in peace microfarms.

Chapter 8. What are Peace Microfarms?

Envision 10 million Green Masterminds growing food and other valuable coproducts in peace microfarms globally. Imagine what each grower can do for the health and vitality of their family and community.

Freedom foods grow in peace microfarms that save precious natural resources for our future generations. Peace microfarms offer a novel approach to avoid conflicts over the fossil resources currently used to produce food. Planet Enriching Algae Cultivation Ecosystems, (peace) microfarms use abundance methods that enable growers distributed globally to recycle nutrients and energy from sterilized waste streams.

Today, most people cannot grow food locally because they lack the weather, fossil resources and money required for crop inputs. When developed, peace microfarms will allow individuals and communities globally to use affordable inputs to grow foods locally. Peace microfarms grow food and other forms of energy sustainably independent of climate, altitude, latitude, geography or politics.[127]

Please follow our progress in microfarm R&D and contribute your ideas for designing effective microfarms at www.AlgaeFuture.org.

Climate Independent Foods

Peace microfarms are adaptable microcrop platforms that enable growers to use low cost inputs to cultivate a wide variety of high value products. Microfarmers practice abundance and use green solar, sunshine, for energy. They may recycle organic inputs from farm, gardens, kitchens or other waste streams that are surplus, low-cost or free. Growers cultivate microorganisms such as algae and the microflora algae attract to produce food for people, feed for fish, fowl, dairy, and meat animals. Other growers grow and flow their culture to produce rich organic biofertilizer for gardens or fields.

Growers practicing abundance are essentially green solar gardeners as they transform solar energy to rich, nutritious plant biomass. The green biomass concentrates energy in chemical bonds that are portable and may be used directly for food or transformed to many other forms of energy. Microfarms grow naturally biodiverse microcrops, so no genetically engineered seeds are needed.

Microfarmers use four configurations, Table 8.1. Estimated yield is compared with field crop protein production such as corn.

Table 8.1 Microfarm Configurations

Microfarm configuration	Estimated yield
1. Open pond or raceway	10 times
2. Covered pond or raceway	20 times
3. Semi-closed culture	25 times
4. Closed or controlled environment	30+ times

Covered microfarms allow growers to extend the season. Semi-closed and closed systems allow year-round, independent of weather. Microfarmers train indigenous, local algae to produce proteins, oils, carbohydrates and other coproducts rapidly. Some growers will cultivate exotic species from algae collections but locally adapted species typically outperform others. A peace microfarm graphic in Figure 8.1 shows how low cost inputs are transformed into a variety of high value products.

What are Peace Microfarms?

Figure 8.1 Peace Microfarm

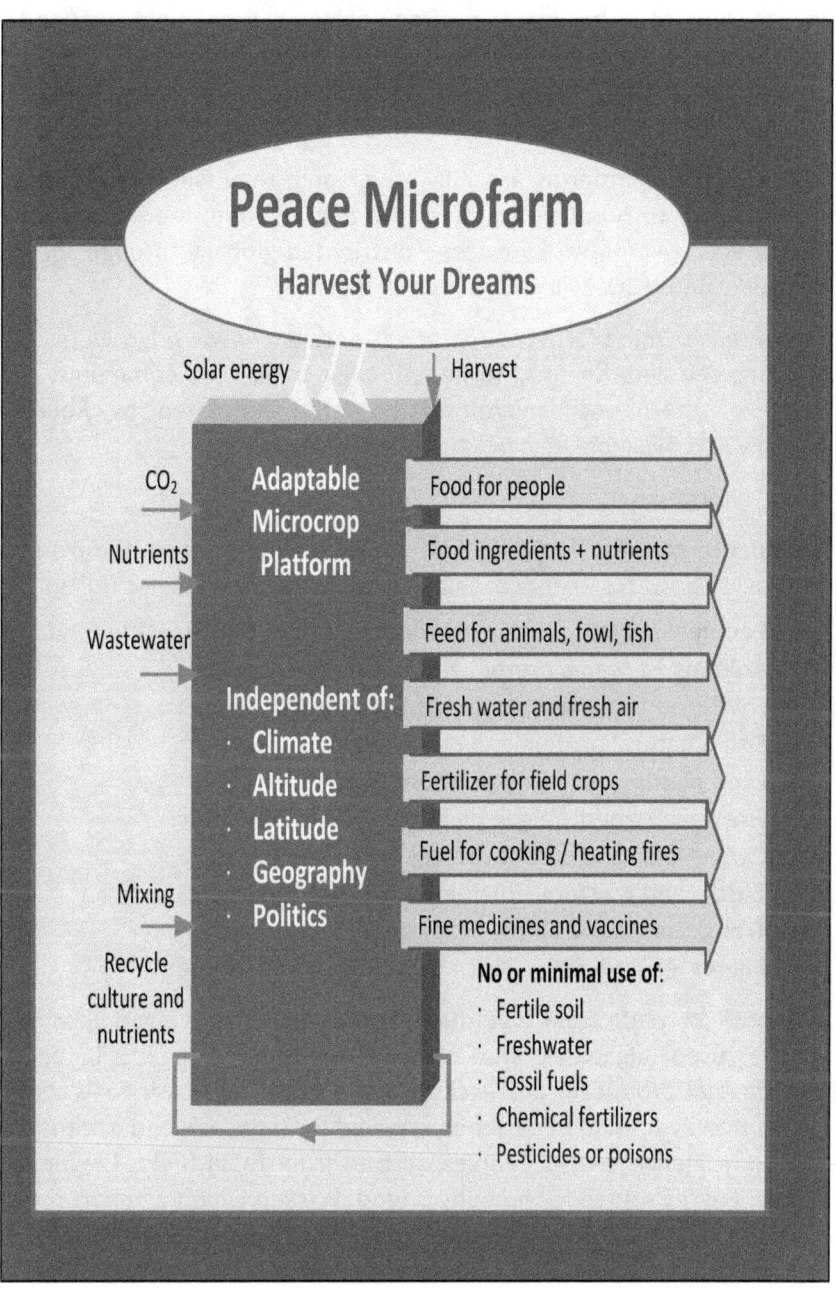

Climate Independent Foods

Microfarms can be sited in nearly any location. The footprint may fit in a corner of a backyard, rooftop, balcony, barn, field, wetland, desert, prairie or other non-crop land. Microfarms may serve a family, community garden, village, coop, community or city. Locations with insufficient sunshine may use efficient LED grow lights for energy.

When peace microfarms are fully developed, they will give growers the freedom to produce a natural diversity of food, feed and other coproducts. Microfarm knowledge distributed globally through social networks and NGOs will create food democracy.

The French microfarm cooperative network allows growers to produce spirulina for food to benefit their family and community. A video of the French microform cooperative created by Robert Henrikson is available at www.AlgaeCompetition.com.

Algae energy

Abundance growers cultivate the fastest-growing plant on the planet to provide portable energy usable in a multitude of ways, including:

- **People** – organic protein, nutrients and micronutrients in food.
- **Animals** – organic protein and nutrients in fodder.
- **Fowl** – natural protein and nutrients for birds.
- **Fish** – natural protein and nutrients in fish feed.
- **Land plants** – rich, full spectrum organic fertilizer.
- **Fire** – high-energy algae oil for cooking and heating.
- **Cars** – lipids and carbohydrates refined to biofuels.
- **Trucks and tractors** – high-energy clean, green diesel.
- **Trains, boats and ships** – high-energy clean diesel.
- **Planes** – high-energy, clean aviation gas and jet fuel.

Any product made from fossil fuels can be made from algae because nature used algae as the primary feedstock for fossil fuels. Commercial producers are excited about replacing fossil fuels with algae. However, human societies survived for many millennia without the convenience energy sources derived from fossil fuels. The most critical energy source for humans is food. We survive only a short time when deprived of the vital energy supplied by food.

What are Peace Microfarms?

Growers cultivate microbial communities, which may be pure strains of algae, but are often diverse communities of algae and the multitude other microorganisms algae attract. Microflora communities thrive in aquatic and moist terrestrial settings and include algae, fungi, bacteria, viruses, slimes and other tiny organisms. This diverse array of microorganisms works symbiotically to produce compounds valuable to plants, animals, fish and humans.

Industrial or organic farmers may use abundance methods to recover and reuse the energy and nutrients in the farm waste stream to reduce production costs while improving soil fertility; crop yields, and produce taste, nutrition and quality. Urban gardeners may source nutrients from municipal and industrial waste streams to grow rich algae biofertilizers that speed plant growth and development as well as increase produce size, taste, texture, color, nutrition and quality.

Abundance provides substantial value for farmers by avoiding fossil inputs, increasing productivity, and reducing costs. Microfarmers can improve soils and reduce water, energy and fertilizer waste while decreasing soil erosion and air, water, and soil pollution, Table 8.1.

Table 8.1 Abundance Benefits for Smart Microfarmers

Benefit	Crop inputs and costs
Fossil inputs	Eliminate or minimize scarce and expensive fossil inputs including fertile soil, fresh water, fossil fuels, fertilizers, and fossil agricultural chemicals.
Plentiful inputs	Produce food, coproducts and other forms of energy using solar energy, CO_2, and wastewater.
Improve yields	Enhance yields of protein, lipids, carbohydrates, energy and other target compounds 20 to 50 times per unit of land each year.

Climate Independent Foods

Crop diversity and nutrition	Expand crop diversity, providing better nutrition, micronutrients, vitamins, and minerals.
Decrease production risk	Harvest biomass daily to remove or diminish production risk.
Expand geography	Enable food and energy production anywhere on earth with sunshine.
Embrace global warming and climate chaos	Crops produce effectively despite prolonged heat, droughts, more extreme storms, salt invasion, rising oceans, wild fires, and pest infestations.
Enable the poor and malnourished	All people can produce food when the inputs are free or surplus, assuming they have access to growing systems and sufficient training.
Nutrient recovery and delivery	
Waste stream	Transform a cost, getting rid of agricultural wastes, to a profit center. Algae use solar energy to recover nutrients.
Bioavailable delivery	Deliver nutrients at the right growing cycle stage in bioavailable form that plants can use.
Boost yields for field crop farmers	
Texture and taste	Improve produce texture and taste through the immediate bioavailability of micronutrients.
Productivity	Improve crop yield, speed to maturity, size, weight and quality 30-50% by providing bioavailable nutrients.

What are Peace Microfarms?

Vitamins and minerals	Enhance the presence, quality and availability of vitamins and minerals 20-30% in produce with bioavailable nutrient and micronutrient delivery.
Digestible nutrients	Increase the presence of digestible nutrients in produce 30% with organic biofertilizers.
Regenerate soils	
Soil compaction	Reduce soil compaction and increase prosody 500% to stimulate root growth, make room for microflora and worms to enhance plant strength.
Crust	Strengthen the soil crust to add nutrients, organic material, and minimize erosion.
Soil structure	Improve topsoil structure by expanding the humus and organic material in the soil.
Soil microbes	Use algae to attract microbial communities that act to enhance crop health and productivity.
Soil moisture retention	Improve soil moisture retention and decrease heat and drought stress.
Improve agroecology	
Fertilizer pollution	Reduce air, soil and water pollution by using fewer chemical fertilizers.
Erosion	Minimize soil loss to wind and water.
Agricultural chemicals	Minimize pollution from agricultural poisons by diminishing or eliminating them.
Bioavailable	Deliver bioavailable nutrients to the soil

Climate Independent Foods

nutrients	precisely when plants most need them.
Greenhouse gases	Reduce greenhouse gas emissions, especially CO_2, methane and nitrogen oxides.
Tillage	Reduce the need for tillage and soil disruption.
Organic farming	Support and accelerate the transformation from industrial farming to organic farming.

Microfarmers cultivate algae and possibly other microorganisms as they follow one or a combination of four Sustainable and Affordable Food and Energy, (SAFE) production paths, Figure 8.2. Peace microfarms support all four pathways.

Figure 8.2 Abundance SAFE Production Paths

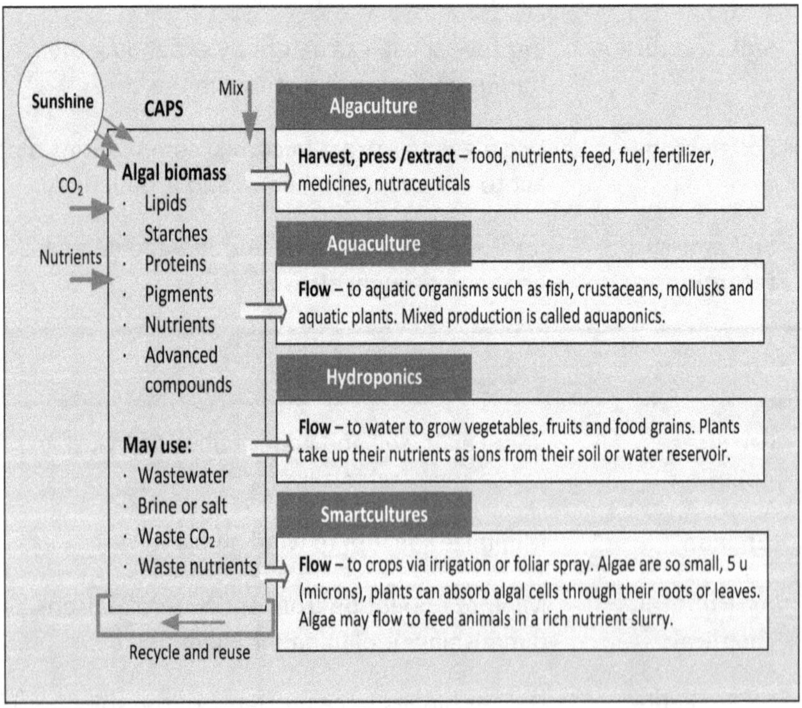

Inputs: Sunshine, CO_2, Nutrients → CAPS

Mix → Algal biomass:
- Lipids
- Starches
- Proteins
- Pigments
- Nutrients
- Advanced compounds

May use:
- Wastewater
- Brine or salt
- Waste CO_2
- Waste nutrients

Recycle and reuse

Algaculture — Harvest, press /extract – food, nutrients, feed, fuel, fertilizer, medicines, nutraceuticals

Aquaculture — Flow – to aquatic organisms such as fish, crustaceans, mollusks and aquatic plants. Mixed production is called aquaponics.

Hydroponics — Flow – to water to grow vegetables, fruits and food grains. Plants take up their nutrients as ions from their soil or water reservoir.

Smartcultures — Flow – to crops via irrigation or foliar spray. Algae are so small, 5 u (microns), plants can absorb algal cells through their roots or leaves. Algae may flow to feed animals in a rich nutrient slurry.

What are Peace Microfarms?

Algaculture

Algaculture grows microalgae or macroalgae, (seaweed) for commercial purposes or domestic needs. Extraction of the algae biomass enables the farmer to use the energy, nutrients and a wide spectrum of valuable coproducts, Table 8.2.

About 33% of the algae grown commercially today feeds fish and shellfish. Nori, used to wrap sushi, leads the sea vegetable market and exceeds $3 billion globally. Spirulina leads the microalgae market with about 5000 metrics tons a year. Most spirulina is currently sold as a health supplement to provide essential micronutrients.

Algaculture producers use many microfarm shapes, sizes and forms including ponds, troughs, semi-closed and closed systems. Food, health food, feed and nutraceutical producers include Earthrise, Phyco Biosciences, Algae Biosciences, Aurora Algae, Solazyme, Seambiotic, Cellana, and Martek Biosciences.

Microfarms sited near a carbon source such as a waste pile on a farm, coal fired power, cement or manufacturing plant gain the advantage of a free carbon source. Every ton of algae consumes nearly two tons of CO_2, so free carbon reduces operational costs.

Other microfarms are sited near a wastewater treatment facility to gain access to free nutrients. Some growers may source organic wastes from farm, garden or other waste streams. Algae grow well in fresh water but communities have competing needs for sweet water. Many communities have substantial sources of gray brackish or wastewater that is not potable but excellent for growing algae. Some farms have reservoirs, ponds or wetlands that capture farm runoff, which are perfect for growing algae.

Half of the water stored in the earth's crust is brine water, which is too salty for human use or for irrigation. Algae thrive on brine water, which often carries the full spectrum of essential nutrients. Many deserts, including in the U.S. Southwest, have huge underlying oceans of brine water in relatively shallow aquifers. These brine aquifers could produce millions of tons of algae biomass for 400 years.

Climate Independent Foods

Table 8.2 Algae for Food, Biofuels and Novel Solutions

Food	Biofuels	Novel Solutions
Primary • Protein • Lipids – oils • Carbohydrates • Nucleic acids **Secondary** • Flour • Meat enhancer • Ice cream • Milk substitute • Sugar substitute • Sea vegetables • Food ingredients • Emulsifiers and thickeners • Novel flavors and textures • Pigments • Health foods • Nutraceuticals • Omega 3s **Feed and fodder** • Pets, fish, fowl • Meat animals • Micronutrients • Medicines and vaccines	**Primary** • Gasoline • Clean diesel • Methanol/ethanol • JP-8 jet fuel **Secondary** • Aviation gasoline • Alcohols • Hydrogen • Asphalt • Plastics, biodegradable • Rubber substitute **Biofertilizers** • Organic N-P-K • Bioavailable target nutrients • Micronutrients • Plant hormones • Soil organics • Build soil structure • Improve porosity • Plant growth regulators • Natural pesticides • Natural herbicides	**Air** • Carbon sequestration • Carbon capture/recycle • Capture sulfur • Capture heavy metals **Water – clean** • Waste streams – municipal, industrial, farm, brine and ocean • Recover heavy metals **Cosmetics** • Moisturizers • Skin care **Local algae production** • Foreign aid • Disaster relief • Hunger and poverty **Medicines** • HIV / AIDS and SARS • Vaccines • Antibiotics /antiviral • Burns and bruises • Stomach remedies • Anti-cancer toxins • Pharmaceuticals • Advanced compounds

What are Peace Microfarms?

Growing algae as fodder for animals, birds or aquatic creatures will be popular in many settings because animal feed requires lower levels of cleanliness, (except for pets in the U.S.), than producing food for direct human consumption. Pet and animal foods that enter the human food chain have high quality requirements, similar to human foods.

Wastewater microfarms can produce food quality algae with the proper safeguards in place. Many human-grade valuable coproducts may be extracted from wastewater algae such as vitamins, minerals, antioxidants, trace elements pigments, oil and carrageen.

Hydroponics

Algaculture producers may produce a slurry or solid product similar to fish fertilizer for use in hydroponics. Farmers can grow algae next to their hydroponics unit and a portion of the algaculture flow to containers where vegetables, grains and fruits grow in the rich algae water. Algae provide all of the macro and micronutrients necessary to grow large, colorful and tasty produce.

Grow algae to feed vegetables in water – hydroponics

Hydroponic farmers grow plants using mineral solutions in water rather than soil. In natural conditions, soil acts as a mineral nutrient reservoir but the soil itself is not essential for plant growth. Terrestrial plants grow well with their roots in an inert medium such as perlite, gravel, mineral wool or nutrient solution. Hydroponic crop yields are be no better than crop field yields with good soil. Crop yields are limited by factors other than mineral nutrients; especially light. Hydroponics growers produced vegetables on Pacific volcanic islands that lacked fertile soil in World War II. Hydroponics saved considerable transportation cost during the war. Similar growing systems are used today to feed scientists in Antarctica.

Aeroponics – Fine mist systems

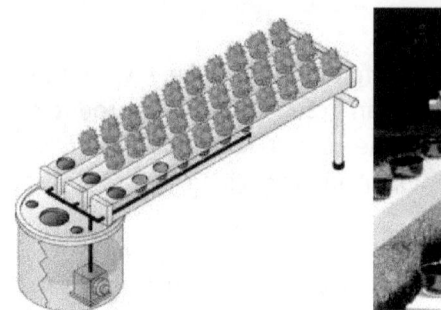

Aeroponics, developed largely by NASA for space travel, grows plants in an air or fine mist environment without soil or aggregate medium. Aeroponics culture differs from both hydroponics and *in-vitro* production (plant tissue culture). Unlike hydroponics that uses water as growing medium and essential minerals to sustain plant growth, aeroponics cultures grow without an aggregate medium. Growers transmit nutrients by water mist, so aeroponics is actually a form of hydroponics.

Aquaculture and Aquaponics

Aquaculture farmers grow fish and shellfish that feed on aquatic plants such as algae. Algae represent the preferred diet for most fish fry (immature fish) because the cells are small enough for the fry to eat. Most fish evolved on an algae diet in their natural settings. Most fish grow faster and have fewer digestive problems on algae compared with food grains.

The Chinese have practiced aquaculture since 2500 BC. Today, half the world's commercial fish and shellfish production comes from aquaculture. A recent scientific study reported that over 90% of the large fish have been extracted from the oceans. Unfortunately, fishermen overharvest many of the smaller fish too, depleting the

What are Peace Microfarms?

food chain. With diminishing natural fisheries in oceans, rivers, lakes and estuaries, aquaculture will play a larger role in our food supply.

Growing algae to feed fish – Aquaculture

Aquaponics integrates fish and plant farming. Farmers grow algae to feed fish that add urea to the water. The nitrogen rich water flows to hydroponic greenhouses where vegetables and fruits grow in the high nutrient water. Polycultures can grow food with renewable energy and in closed systems, minimize consumption of fossil resources, including power and fresh water.

Smartcultures

Sustainable Micro Algae Regenerative Technologies, (smartcultures), enable field crop farmers to recover, recycle and reuse the energy and nutrients in their farm's waste stream to improve crop quality, taste and yields, while reducing operational costs.[128]

Smartculture farmers skip the harvest step and simply "grow and flow" the algae culture directly to their fields to recycle organic fertilizer to their crops that is immediately bioavailable to the plants. Smartcultures deliver 74 nutrients and trace elements that plants use to grow faster and stronger and produce higher yields.

Smartcultures employ a set of technologies that imitate nature to provide enhanced foundation (soil structure) and food (nutrients) to plants. Every farmer and gardener knows plants thrive in amended soils; they grow faster, stronger, and larger, and they have better taste and texture. Animal farmers can recover most of the energy and nutrients remaining in the farm waste stream and recycle it to feed farm, dairy, poultry or meat animals

Smartcultures begin at the crop foundation – soil – with tiny microflora attracted by algae in plant roots that are ingeniously self-regulating and self-regenerative. Smartcultures move farmers toward abundance production by significantly reducing, but not eliminating the use of fossil resources for growing field crops. Farmers using smartcultures are able to leave every field better than they found it.

Field crop farmers using smartcultures can use 80 to 100% fewer agricultural chemicals because algae biofertilizers provide growth hormones that make plants stronger and able to produce natural pest and disease defenses. Growers can reduce soil compaction 500%, enabling significantly longer and stronger root structure. Stronger roots give plants a deeper reach for nutrients and soil moisture. Healthier plants on a stronger foundation need less water and are less vulnerable to weather, winds, weeds, disease and pests.

Algae deliver biofertilizers to field crops through irrigation or spraying. The algae continue to grow in the field, increasing valuable organic material, humus, and improving moisture retention. and soil fertility.

Algae growing in a Turf Scrubber and Air Dried

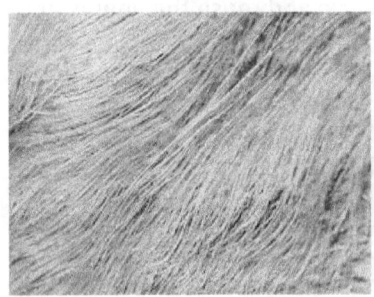 

Algae can remove nitrogen and phosphorus in livestock manure runoff. Walter Mulbry, an Agricultural Research Service scientist set up four algae turf scrubbers outside dairy barns. The shallow 100-foot raceways are covered with nylon netting that creates a scaffold where the algae grow. For the next three years, from April until December, a submerged water pump at one end of the raceways circulated a mix

What are Peace Microfarms?

of fresh water and raw or anaerobically digested dairy manure effluent over the algae. The raceways supported thriving colonies of green filamentous algae.[129] Algae grown in open raceways create diverse microbial communities of microalgae and other microflora.

Each acre of algae raceway removed nutrients from 20 cows' manure. The system recovered 60-90% of the nitrogen and 70-100% of the phosphorus from manure effluent.

The research team calculated the recovery cost was comparable to other manure management practices—with a per pound cost of $5 for N and $25 for P. Dried algae made excellent organic fertilizer as corn and cucumber seedlings grown in algae-amended potting mixes performed as well as those grown with commercial fertilizers.

Field trials

Two years of field research by the first author with a multinational food production company shows smartcultures methods can:

- Increase income from 20 to 50% by improving crop yield and quality – micronutrients, vitamins, antioxidants, color, taste, texture and shelf life.
- Lower fuel consumption by 20 to 30%.
- Decrease chemical fertilizers by 30 to 50%.
- Reduce air, soil and water pollution by 80 to 95%.

Farms without irrigation systems spray the algae solution on the fields. The algae not only provide an organic fertilizer delivery system directly to the roots of crops but algae continues to grow in the field, as long as moisture is present, regenerating soils and creating additional organic material. Algae's ability to extract in situ nutrients provides a tremendous advantage. Most farmers have available waste streams from human, animal and vegetative wastes on which algae can thrive. Rather than spending 30-40% of their production costs on fertilizer, algae may cut nutrient cost in half because algae can recover 90% of the nutrients from the farm waste stream. The typical farm waste stream contains about half of the nutrients needed for the next crop. Therefore, the net fertilizer reduction approaches 50%.

Smartcultures grow algae in the farm waste stream to fertilize fields

More microfarm examples are available at AlgaeFuture.org.

Smartcultures transform agricultural methods and rather than mining and paying high prices for chemical fertilizers and using them once, farmers can continuously recycle nutrients. Rather than systemically extracting soil nutrients and organics, farmers can cultivate algae and microflora to add nutrients and organics to their fields. Rather than using chemicals that destroy soil microbes and soil structure, smartcultures cultivate microbial communities that improve soil structure. Industrial agriculture degrades soil, promotes erosion and creates severe pollution, while smartcultures improve soil structure and reduce nutrient waste, erosion and pollution.

Recent fertilizer price escalation has pushed the cost of fertilizer up to 35% of farm operating costs. Farmers face two other serious problems with industrial fertilizers: bioavailability and erosion. Microorganisms in the soil, e.g. algae, must first break down chemical fertilizers before the plant can absorb them. The process may take months or years. Consequently, farmers have to put on far more fertilizers than the plant needs in order to maximize growth. Much of the applied fertilizer does not reach the crop and erodes with irrigation, winds and rains. The next year, the farmer must apply even

What are Peace Microfarms?

more fertilizer to achieve the same yields. This model is sustainable only as long as fertilizers are cheap and soils do not wear out.

Farmers can improve the quality and quantity of field crops because algae biofertilizers are immediately bioavailable to the plants and create almost no waste. Some algae are able to unlock nutrients in the soil, such as phosphorus. Over 95% of the P in some soils is locked in large molecules that are not absorbable by plants. Algae can solubilize the P and other elements, making them bioavailable.

Smartcultures can deliver precise amounts of target nutrients carried in algae biofertilizers at specific times during a crop's growing cycle – which can maximize germination, early growth, maturation and fruiting. Microfarms near fields can overload one or more nutrients and deliver them to crops exactly when needed. For example, adding more calcium when the crop is fruiting may enhance fruit size, weight, color, texture and taste.

Farmers can save money and energy by lowering their consumption of fuel due to easier cultivation. In some settings, smartcultures have improved soil porosity (loosen compacted soil) by 500%. Smartcultures reduce air, soil, and water and pollution because algae biofertilizers and plant growth hormones significantly diminish the need for agricultural chemicals.

Drip irrigation can deliver algae biofertilizers precisely to the roots, minimizing water use and the waste of nutrients. Algae continue to grow in the soil while moisture is present, which adds rich organic matter and conditions the soil, making it more erosion resistant. This model may also use no or minimal till to minimize soil disruption and provide longevity to the water-efficient drip irrigation system. In non-irrigated settings, growers may apply the algae culture with a field or aerial sprayer.

The seed catalogue, *Cook's Garden* sells a seaweed product called Sea Magic.[130] The algae fertilizer contains cytokinins and 17 amino acids that encourage stronger, lusher growth, more sugar production in fruiting crops, and increased blooms. The company claims this seaweed fertilizer increases yields, performance and flavor. Our field and garden research supports their claims, which include 24% more tomatoes, 25% more grapes, 34% more cucumbers, and 47% more peppers.[131] Our peppers were closer to 25% yield improvement, but our garden already benefits from excellent organic soil with microorganism communities. Most nurseries and seed catalogues sell multiple algae fertilizer products made from mined fossilized seaweed. Similar seaweed fertilizers are plentiful in hydroponic supply stores and claim similar yield boost.

Other forms of food production

Human societies will need every sustainable form of food production to meet global food demand, including organic production, controlled environment agriculture and vertical farming.

Organic production offers health advantages for consumers and producers. However, organic production is not sustainable because it consumes more cropland, freshwater and fuel than industrial agriculture. Organic growers use less inorganic fertilizer and agricultural chemicals, which is better for ecosystems. Organic growers often produce local to consumers, consume few agricultural chemicals and create substantially less ecological pollution than industrial agriculture.

Unfortunately, most countries, including the U.S., have insufficient cropland to grow the biomass needed for composting organic growing systems. Even if sufficient compost was available, the energy and time costs involved in making compost and cultivating the material into fields precludes widespread adoption.

Controlled environment agriculture and vertical farms offer the promise of resource efficiency and pollution free food. Capital and operating costs may be too high for widespread use, unless policy leaders tax food distribution miles. Vertical farms have been proposed for decades but production costs undermine profitability and viability. Time and testing will demonstrate methods of sustainable production.

Chapter 9. Microfarm Examples

God Almighty first planted a garden. And indeed, it is the purest of human pleasures.
— *Sir Francis Bacon*

A few small-scale algae cultivation systems are available currently. Several companies will provide scalable algae cultivation kits in the near future. Current systems require considerable attention, knowledge and labor. The best solution will include growing support that minimizes the learning curve and maximizes grower success.

Home growers have not cultivated algae successfully, except for spirulina. Cultivating algae successfully today requires considerable knowledge, experience and technology. The industry needs a solution that empowers gardeners and farmers to grow algae on a small scale and then migrate to larger cultivation systems.

The key innovation will be the addition of remote monitoring capability to growing systems. Culture information will upload through the Internet to a laboratory staffed with experts that provide advice to growers. Eventually, expert systems will provide grower advice in real time, which will greatly simplify algae cultivation.

Climate Independent Foods

About 150 commercial algae producers cultivate algae in 2012, primarily for special compounds such as omega-3 fatty acids, nutraceuticals, and health foods. Firm profiles are available at the Algae Biomass Organization and *Algae Industry Magazine*.

Home systems

Several providers offer algae kits that enable home growers to produce small amounts of algae.

AlgaeLab.org

Aaron Wolf Braun founded algaelab.org and teaches workshops on how to grow algae. Participants complete the seminar with a take-home growing kit similar to the picture. Aaron, a.k.a. Dr. Friendly, is working on a book on growing algae that includes science trends and technology, and aquaculture from micro to large scale.

The AlgaeLab store offers the basic supplies needed to grow spirulina at home. The AlgaeLab.org site has pictures of the cultivation and harvest processes.

Home cultivation systems currently produce primarily the blue-green algae spirulina because it is easiest to grow and harvest. Spirulina has many advantages, including 70% protein, high-density micronutrients, antioxidants, vitamins, minerals and trace elements.

One of the most common requests people make is "How can I grow my own algae?" Jean-Paul Jourdan published manuals and curriculum on how to grow spirulina, encouraging many more people to get

Microfarm Examples

personally involved in growing algae locally for family, friends and community. Algae microfarms fit with the growing Do-It-Yourself movement. Microfarms align with the trend for growing food and herbs indoors, in greenhouses, on rooftops and in empty lots urban, backyard and community gardens.

Smart Microfarms

Robert Henrikson, CEO of SmartMicrofarms.com and AlgaeFuture.org, grows spirulina in several scalable indoor, porch and backyard units. He harvests up to a pound a day and his family and friends eat about half fresh. The rest is frozen to cubes or dehydrated.

French development of appropriate scale low cost technology has been ongoing in India by agencies of the government with NGOs. In Africa, humanitarian groups from France have built village projects, including Dr. Ripley Fox and Antenna Technologies.

Appropriate scale village farms in the developing world have led to the emerging movement of commercial spirulina algae microfarms. Growers produce for their families and local markets. Many of the charitable organizations diffusing algae production originated from France. Over 100 algaepreneurs produce algae locally in France today and over 500 are expected within five years. The algae microfarm

Climate Independent Foods

movement that began in France has recently migrated to Spain and shows signs of coming to North America.

Spirulina producers have formed a cooperative to learn from one another in Southern France to as far north as Normandy. The school at the CFPPA Center in Hyères trains growers. In 2010, growers established the Fédération des Spiruliniers de France and developed a Charter of Good Business practices.

Laurent Lecesve, at his Eco-Domaine farm in Normandy, France.

Cultivation in ponds

Algae pond systems in the U.S. were first developed for wastewater treatment. Producers recovered the biomass, converted it to methane, and burned it as a local source of energy.[132] When fuel was cheap, the energy value from algae was considered incidental.

Wastewater Treatment and an AquaFlow Pond in New Zealand

Algae wastewater treatment offers a low energy and low cost means for cleaning polluted water. Proven technologies kill parasites and pathogens in the wastewater and algae remove the substantial organic material. Producers can recover the algae biomass for use in animal feed, fertilizer, green chemicals and advanced compounds.

Microfarm Examples

Some producers use the algae oil for green energy and the residual protein for animal feed.

A variety of other algae growing systems are available. In open ponds, the productivity ranges from 10-35 grams/meter^2/day (36-128 metric tons/hectare/year) on a dry weight basis.

Algae Grown in Furrows and an open Raceway

Algae ponds typically are shallow, six inches to three feet deep, in order to maximize cell access to light. Algae grow quickly and new cells shade older cells. Unmixed ponds have growth only in the top two inches of the water column. Ponds are mixed with a paddle wheel or compressed air that keeps the culture moving around the raceway and up and down the water column. Water movement needs to create sufficient turbulence to move cells to the surface so they can absorb photons. Large ponds used for municipal water remediation typically bubble a mixture of CO_2 and air to move the water.

Size affects water circulation, operating costs, mixing systems and species selection. Mixing gives cells access to light, prevents cells from settling to the bottom and avoids thermal and oxygen stratification in the pond. Effective mixing increases cell density, which reduces harvesting cost. Harvesting typically occurs with microscreen, centrifuge, filters or flocculation.

Algae are currently being grown outdoors in ponds, greenhouses, photo bioreactors, fermenters and hybrid systems combining bioreactors and ponds. As new technologies and systems arrive, algae microfarming will be less costly, easier and more accessible for more people globally. Emerging technologies will stimulate a distributed model of scalable growing systems. Community microfarms will grow algae for food, feed and biomedicines for the nearby region.

Raceways scale any size and are constructed as a loop. Raceways have the advantage of simple, low cost construction and maintenance. Most algae production today occurs in open raceways because raceways are cheap to build and operate. Some algae, such as *dunaliella* are grown in deep saline ponds with little mixing. Ponds are most productive in tropical, subtropical and temperate areas with warm temperatures, low rainfall and little cloud cover.

Disadvantages of ponds include lower productivity due to lack of temperature control and water loss from evaporation. An open pond loses about as much water due to evaporation as a grain field consumes in irrigation. Water loss increases retained salts and impacts culture stability. Some open ponds use seawater, waste or brine water, which often makes the water free but does not slow evaporation or salt concentration.

Algae Ponds and a Plastic Bag

Outdoor ponds make it difficult to control algae predators such as amoeba, ciliates, bacteria, rotifers, viruses, fungi, and zooplankton that can decimate the algae biomass within hours.[133] Open ponds are vulnerable to contamination from dust, windborne organisms, insects, and birds.

Commercial algae producers have devised strategies to minimize contamination by opportunistic weed algae in open ponds. Producers grow *spirulina* at high bicarbonate concentrations with high pH. *Dunaliella* grow in high saline water to discourage competing species. To manage contamination for *chlorella*, producers grow the biomass in batches with increasing volumes. Growers harvest the entire batch and the ponds purged and cleaned. Then new batches are restarted from clean laboratory cultures.

New research centers such as AzCATI at Arizona State University and algae incubators are testing various pond and photobioreactor systems to compare results and develop smarter automated systems. The pictures show two open raceways at the AzCATI test bed at Arizona State University.

Raceways at AzCATI built by Nano Voltaix

Most microalgae need light and carbon dioxide but they vary substantially by specie in nutrient and environmental requirements. Some species grow well in unlined ponds in Australia but the same variety may not flourish in unlined ponds in India or China. Local conditions often dictate the design and construction of open ponds as well as species selection and production methods.

The significant drawbacks of the open raceways have prompted the development of closed systems, called photobioreactors, made of transparent tubes or containers in which the culture is mixed by either a pump or CO_2 and air bubbling.

Closed microfarms

Covered, semi-closed or closed containers are designed to capture maximum solar energy. Systems may vary in size from a several square yards to several acres. Growing containers provide considerable visual variety and may be covered ponds, plastic bags, plastic sheets, flat plates, tubes or glass – anything that allows light to penetrate.[134] Some indoor systems use fiber optics or mirrors to capture sunlight or to add artificial light. Closed systems are more capital intensive than outdoor ponds.

Closed systems minimize contamination, permitting the cultivation of a single microalgae species. Closed systems offer better control over biocultural conditions such as pH, light intensity, carbon dioxide, nutrients and temperature. Tighter control lowers CO_2 losses due to out-gassing and minimizes evaporation.

Closed systems are often two to five times more productive per unit area than open microfarms. Closed systems offer the significant advantage of weather independence, with year round production. Growers counter cold weather and low light by adding sources of heat and light. Closed systems offer tighter control of contamination from unwanted algae, zooplankton predators, dust and debris. Closed systems may add cooling costs to guard against overheating in hot climates and higher cleaning costs from fouling.

Consumer research indicates creates negative consumer perceptions about the term bioreactor. Microfarm provides a constructive alternative. Photobioreactor implies that the sun excites plant cells to produce biomass through photosynthesis but naïve observer's associate reactors with nuclear power. Additionally, the term bioreactor has become synonymous with garbage waste disposal.[135] Consequently, the preferred terms are microfarm, biofactory or cultivated algaculture production system, CAPS.

Closed microfarm producers select algae species that maximize the characteristics desired such as biomass percentage of lipids, protein, or component product. Producers of food select to maximize biomass protein while nutraceutical growers select algae species with high lipid content. Growers have a new tool developed at Michigan State University to simulate growing conditions before investing in an algae growing system.

Phenometrics developed the computer controlled Environmental Photo BioReactor (ePBR™) for laboratories that simulates realistic environmental conditions such as temperature, light intensity and

Microfarm Examples

CO_2. The unit allows scientists to study algae under the same conditions found in various growing settings.

An array of Phenometrics ePBRs in a Lab

Each ePBR is a measurement instrument, quantifying growth rates, pH, and other factors in the algae culture, which are displayed on the computer monitor in a graph or data display. It is customizable to accommodate various probes and sensors that monitor algae culture and growth metrics. The ePBR™ provides reproducibility, a key factor in algae research experiments.

In between laboratory simulators and field growing systems are laboratory systems that scale.

Professor Joel Cuello developed the accordion PBR in the Biosystems Engineering Laboratory at the University of Arizona. The accordion is a set of transparent bags that scale up to nearly any size to optimize culture productivity. The accordion applies biochemical and ecological strategies for growing algae quickly, wastewater treatment and growing algae fertilizer for hydroponics.

Climate Independent Foods

The term "closed system" is a misnomer because algae predators and weed species invade any growing system. It may be better to think of a closed system as an arrangement that gives the grower more control over production parameters and more but not perfect control over contamination from invasive algae and predators. Growers typically control predators and weed species with a combination of parameters including pH, temperature, salinity and nutrients.

Tubular Microfarms

Ultraviolet (UV)-stabilized acrylic is typically used for construction because compared with glass; it is cheaper, stronger, lighter, more flexible, and easier to fabricate. Assembly of microfarms requires the integration of the various mixing, monitoring, and controlling subsystems.

Flat Plates built by Nano Voltaix for Arizona State University

Artificial lighting

Indoor growing systems with artificial lights allow growers to produce algae in any climate or geography 365 days a year. For example, Algaedyne's algae growing system allows farmers in Minnesota to grow algae on their farms.

Microfarm Examples

Algaedyne Aglae System at St. Cloud State University.

The interior light system produces more food on one acre than field crop farmers can grow on 80 acres. The system uses CO_2 given off by a farm's digester generator, which converts cow manure to methane gas. Manure nutrients are recycled to fertilize algae.

At St. Cloud University, a student team analyzes the algae examines the nutrients recovered. The student team creates economic models for which recoverable algae components, protein, pigments, oils and other components offer the highest market value.

BioVantage Resources offers sustainable bioremediation solutions for industrial, agricultural and municipal wastewater treatment that may employ lighting. BioVantage recovers the nutrients from waste streams with biological solutions and turn the nutrients into high value products. The harvest biomass recovers resources such as phosphorus, metals, protein and energy, or for by-products such as biochemicals, bioplastics, biofuels and/or fertilizer.

BioVantage PBRs with Lighting   **Tank PBRs with Light Pipes**

Climate Independent Foods

The BioVantage integrated high efficiency LED lighting uses optimal and customizable wavelengths for photosynthesis. Light-pipe technology distributes illumination evenly throughout the water column for high-density growth, enabling more algae production within a smaller area. A patented pattern light-pipe illuminates the algae without adding heat so growers do not need heating or cooling. BioVantage's bubble column and tank PBRs can be automated with medium preparation system and full growing system controls.

Algae2Omega developed unique technology for cultivating algae in controlled environments anywhere. Their specialty is converting vacant warehouses to scalable modular production systems that are low-cost, energy-efficient and environmentally sound. The system can grow and algae species in practically and location. Algae2Omega grows algae products for personal care, cosmetics and nutraceuticals.

The tubular PBR shown uses illumination from a LED panel behind the tubes. The solar tubes use solar energy delivered from the warehouse roof with fiber optics.

Algae to Omega Tubular PBR   **Solar Tube**

Vertical, angled or horizontal. Plants are solar collectors and benefit from maximum exposure to the sun. Some angled abundance systems track the sun similar to photovoltaic solar collectors.

Horizontal CAPS, typically flat plate, tubular, or plastic bags, provide several variations to solar exposure. Several companies use plastic bags that are typically rectangular or oval.

Microfarm Examples

Tubular CAPS

Different shapes provide various levels of solar exposure. A wide rectangle, similar to an aquarium, holds a lot of water but does not allow each alga cell to have sun exposure very often. Consequently, thin rectangular tanks, about three inches thick, tend to out-produce tanks that are wider. Tubular tanks may be a few inches wider because they present more surface area around the circumference. Tubular CAPS about six inches wide typically out produce wider tubes.

Experience with various abundance designs in field settings shows the best architecture varies by setting, Table 9.1. The algae biomass often cannot withstand ambient temperatures so they need protection. Without protection such as shading, cultures may be inconsistent in production, become unstable and simply stop growing. Fast growing pure algae cultures do not remain clean indefinitely and weed algae and predators must be removed.

Table 9.1 CAPS Type and Trade-offs

Type	Description	Limitations
Open pond	Economical, easy to manage, good for mass algae cultivation, considerable global experience and shared knowledge base.	Low culture control Stability issues Weak productivity High land and water use Species contamination Predator invasion
Vertical column	High mass transfer, good mixing with low shear stress, low energy consumption, scalable.	Small illumination surface Expensive construction Shear stress problems Cleaning issues.

Flat plate	Large illumination surface, good light path, good biomass productivity, relatively cheap, easy to clean, low oxygen build-up.	Scale-up challenges Culture stability Temperature stability Possible shear stress.
Tubular	Large illumination surface, good light path, relatively cheap.	Gradients of pH, dissolved oxygen and CO_2, tube fouling, high land use if laid flat.

Any algae cultivation system can be covered. Covers protect against predators and weed algae species in open systems. Covers extend the growing season by retaining heat in cold settings and shading in hot conditions. Smart growers flow water through an underground cistern to stabilize culture temperature.

The algae industry continues to experiment with variations in CAPS designs and operations. Most likely, low cost producers of commodities like feeds will use open systems. Applications producing algae oil for biofuels will probably use closed systems to maximize growth speed, vitality and species homogeneity. Maximum total production may be achieved with hybrid systems where closed CAPS grow pure inoculate strains quickly to desired densities that flow to open or covered ponds for production and harvest.

Cost dominates the challenge of algae production. In order to be commercially viable, algae must produce biomass at lower dollar and energy costs than other food and energy alternatives. The National Renewable Energy Laboratory, (NREL), Algal species Program, for example, concluded in 1995 that closed systems were impractical for algae production because they were too expensive to build and maintain. Nearly all algae production to date occurs in open ponds but this will change quickly since most planned algae production in the U.S. intends to use closed or semi-closed production systems.[136]

Microfarm Examples

Fermenters

Most algae species are autotrophs and use solar energy directly to produce organic substrates that store chemical energy from water and CO_2. Many species can also function as heterotrophs and are able to metabolize organic substances to create and store the chemical energy needed for their lifecycle. Heterotrophic algae can be grown in large containers called fermenters without light and are fed sugar as their primary energy source.

Fermenters offer several advantages including considerable published information on production as well as tested commercial growing systems. Heterotrophic growing systems may have lower operational costs than light-based systems, as long as a cheap sugar feedstock exists.[137] Algae biomass production without light usually uses a pure algae strain called an axenic culture that is free of other contaminating organisms. Molecular biologists have recently genetically transformed autotrophic algae that feed on light energy to heterotrophs. These transgenic cells thrive on sugar in the absence of light. The dual algae growth mode enables more flexibility in designing and operating growing systems.

Quality control

Measurements of process quality vary with the goal for the system. Optimizing food production requires substantial monitoring, testing and assurance that the process meets FDA and sometimes, organic food standards. Automated production systems enable quality control checks continuously for all the critical variables. Quality control may include monitoring for:

- **Biomass** density, color, size, structure, and vitality.
- **Water** temperature, pH (acidity), dissolved O_2, and CO_2.
- **Water** quality and dissolved salts and possibly metals,
- **Mixing** velocity and turbidity.
- **Nutrient** availability for all important nutrients,
- **Contamination** from weed algae or predator invasion.

Climate Independent Foods

Measurement of various component parameters needs to occur throughout the growing process as well as harvest, oil extraction, and component separation.

Production targets

Most of the planned production for algae biomass in 2013 targets large algae farms of several hundred acres to produce algae oils used as liquid transportation fuels. Algae farmed for energy promise extremely high return on investment, once technology challenges are solved. Liquid transportation fuels are the primary algae application receiving most public or private financing.

The algae industry will probably develop similar to traditional farming with small, medium, large, and mega-farms. Mega firms will focus on liquid transportation fuels while other growers will cultivate a diverse array of algae products.

Superb graphics for real and imagined microfarms are shown in the highlights from the International Algae Competition, *Imagine our Algae Future,* (Henrikson and Edwards, 2012).

Imagine Our Algae Future

Chapter 10. Why not Freedom Foods Now?

Modern foods seem cheap because neither farmers nor consumers pay for extraction, consumption or pollution.

The first question most people ask about freedom foods is, "Why weren't freedom foods and abundance methods invented before?" The answer: poor math. Industrial foods appear to be cheaper because they do not reflect the true costs. Subsidies, resource depletion, fossil energy security, waste, pollution and health care are ignored, and transferred to our children. Consumers will benefit from full natural resource and lifecycle accounting.

Intellectual property, (IP) has hindered new foods too. Freedom foods represent a "natural process," developed by nature, not by man. Natural processes belong to everyone and are not patentable or listed as intellectual property. Agribusinesses that supply inputs to farmers hire legions of technologists and scientists to develop synthetic compounds. These patented compounds, synthetic poisons, and GE seeds sell to farmers for huge premiums that create wealth for the companies and their executives. Agribusiness advertising promotes this intellectual property and creates a belief in better living and farming through chemistry. Monsanto's Round-Up™ is among the most common words in modern farming.

Climate Independent Foods

Consumers are addicted to highly processed foods, with high sugar, fat, cholesterol and salt but few nutrients. Farmers are addicted to increasingly expensive genetically engineered seeds, synthetic fertilizers, and patented herbicides, pesticides and fungicides. The sad irony is that the synthetic poisons kill the beneficial microbes that nature put in the field to feed and nurture plants. Imagine, paying premiums to kill the organisms that work symbiotically with plants to provide nutrients and plant hormones for their vitality and defense.

Modern farmers have bought into chemical fertilizers because they are cheap and easy to apply. Industrial agriculture systemically extracts macro and micronutrients as well as organics from field soils. Each year, many farmers replace only the three N-P-K macronutrients (nitrogen, phosphorus and potassium). With every crop, micronutrients diminish along with soil organics, which creates nutrient dilution and hidden hunger, which diminishes color, aroma, taste, and texture in produce.

Modern farmers use large, heavy tractors that cultivate quickly but compact soil, which diminishes root growth and accelerates erosion. Farmers buy tons of chemical fertilizers, pesticides, herbicides and fungicides. Crops are developing resistance to chemical fertilizers, so farmers must apply more. Pests and weeds are developing resistance to chemical poisons, which means farmers must use more or change poisons.[138] Plants often absorb less than 5% of the agricultural poisons applied to fields, which creates enormous waste and cost. The residual fertilizers and poisons flow into wetlands, streams and groundwater where they damage and destroy local ecology.

Fertile soil is not an inert medium but a mixture of water, air, minerals and organic matter. In most soils, minerals represent around 45% of the total volume, water and air about 25% each, and organic matter 2-5%.[139] The mineral portion consists of three distinct particle sizes classified as sand, silt or clay.

Soil health depends on the organic component that house many living creatures along with dead material in various stages of decomposition. An acre of living soil may contain 900 pounds of earthworms, 2400 pounds of fungi, 1500 pounds of bacteria, 133 pounds of protozoa, 890 pounds of arthropods and algae, and possibly some small mammals.[140] An acre of soil may contain over 10,000 species of microorganisms, which contributes significantly to the biodiversity in living soil.[141] Unfortunately, industrial agriculture acts to kill the microorganisms with cultivation, soil compaction, chemical fertilizers and agricultural poisons.

Soil organic matter is the smallest but most critical soil component for crops. Soil organic matter interacts to influence soil biological, chemical and physical properties and consists of raw plant residues and microorganisms, (1-10%); active organic traction, (10-40%); and resistant or stable organic matter, (40-60%) called humus.[142] Modern farmers replace the macrofertilizer, but not the humus removed by each crop.

Freedom foods are antithetical to agribusiness firms because foods grown with "natural processes" are not patentable. Similarly, bioavailable algae fertilizers and other plant inputs are natural and are not patentable. Nature was engineering marvelous products that provided for plant needs eons before Monsanto entered the business. Algae and the symbiotic microbes they attract create provide the compounds that enable plants to naturally synthesize many of the advanced compounds they need to grow and to fight disease and pest vectors. Unfortunately, U.S. government farm policy chose to support R&D on industrial food production rather than natural processes.

Farm policy

Food production is dictated by farm policies. Government sponsored research to Land Grant Universities, extension service agents, subsidies, and food support for the hungry are governed by farm policy. The same large agribusinesses that have addicted farmers to their branded synthetic chemistry drive farm policy. Wealthy farmers and agribusiness like ADM, Monsanto and Cargill make enormous political donations to both parties in order to shape policy in commercial agriculture to benefit their interests. Consequently, over 99% of federal grants and R&D go to industrial agriculture. Unsurprisingly, most extension agents who are in place to help farmers and gardeners receive training in industrial production. Less

Climate Independent Foods

than 1% of U.S. federal funding goes to organic production. Government funding for natural growing methods such as abundance and freedom foods rounds to zero.

India and China support natural processes R&D in food production because their leaders realize that fossil resources are finite, increasing in price, and will eventually run out. Both countries have terminated their biofuels programs with food crops for the obvious reason that food-based biofuels drive up the cost of food and the inputs to produce food.[143] China recently put a 135% tariff on their phosphorus fertilizer to insure sufficient supplies for domestic farmers. India's scientists have performed some excellent R&D with natural biofertilizers, especially focused on cyanobacteria that fix nitrogen and reduce the need for nitrogen fertilizer.

As modern farm policies evolved from 1960 to 1990, food supply and sustainability issues were not well articulated. Consumers and political leaders preferred celebrating their brilliance in designing the Green Revolution and the cheap food it provided. Leaders and policy makers ignored critical issues with GE crops, such as the need for additional cultivation, two to three times more irrigation, triple the need for fertilizer and ten times the need for agricultural chemicals.

Few people were aware of nutrition and health issues, food security, fossil resource depletion or global warming before the 1980s. The winds of political rhetoric drowned out the few voices that challenged the fossil foods path such as Prince Charles, Vandana Shiva, Michael Pollan, Miguel Altieri, Alice Waters, and Robert Henrikson.

Today, only one third of Americans believe the scientific consensus that human actions cause global climate change. However, neither politicians nor consumers can deny that humans have caused severe fossil resource depletion and environmental pollutions with our cheap fossil food policies.

Cheap food?

Freedom foods make little sense when the cost consumers pay for industrial foods appear to be so cheap. Appearances can be deceiving.

Why not Freedom Foods Now?

The U.S. government lavishly subsidizes industrial farming, big agribusinesses, big oil, water management and the fossil resources on which food production depends. For example, many farmers pay less than 2% of the true cost of irrigation water – which promotes waste and pollution. Subsidies reduce the real food cost by nearly a third, Figure 10.1. Subsidies are financed with our children's money, in U.S. bonds held by countries like Saudi Arabia, Egypt, Qatar and China.

Figure 10.1 Real Cost of Food

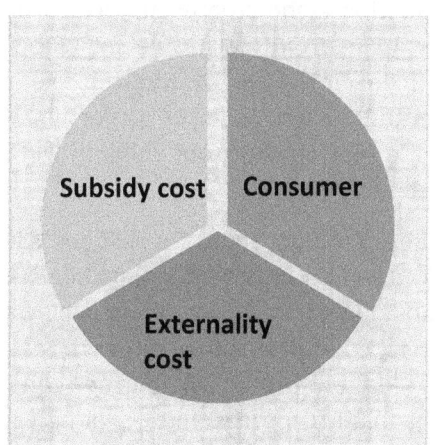

American corn subsidies decimated Haitian farmers because they could not grow food as cheap U.S. food dumped on the country as "food aid." The U.S. corn subsidies also have displaced over a 1.5 million poor Mexican farmers. Farmers were forced to leave their land because they could not compete with subsidized U.S. corn. Many of these farmers added their feet to the flow of illegal immigrants to the U.S. from Mexico. Canada, Mexico and other countries have outstanding lawsuits against U.S. subsidies with the World Trade Organization because these subsidies substantially depress the real price of food grains.

A group of more than 400 agricultural experts, known as the International Assessment of Agricultural Knowledge, Science and Technology for Development concluded through its global and regional studies report that governments and industries need to discontinue environmentally damaging farming methods. At their

Climate Independent Foods

2008 meeting in Johannesburg South Africa, the group recommended, "ending subsidies that encourage unsustainable practices." Political leaders in the U.S. should listen to world opinion because U.S. subsidies amplify resource consumption and pollution. Subsidies today will destroy our ability to grow our own food in the near future.

Another third of the true food cost comes from externalities such as resource depletion, environmental degradation and human health impacts, for which the food supply system fails to account. Environmental degradation alone creates about $45 billion a year in damage. No metrics are currently available for resource depletion. Neither farmers nor consumers pay a nickel for these costs. These hidden costs are shifted to our children. Unfortunately, when the groundwater crashes in the Midwest, our children will not be able to buy sufficient water at any price.

A full lifecycle accounting would show fossil foods are substantially more expensive than freedom food production. Life cycle accounting includes the economic impact of degrading air, water and soil, destroying our fisheries, creating dead zones as well as cost to human and animal quality of life and health. The current generation benefit from over-consuming fossil resources and polluting ecosystems. We ignore resource loss by failing to account for depletion in the price of our food. The next generation will not enjoy the same luxury.

Prince Charles in his Future of Food speech at Georgetown University pointed out the "curiously perverse" economic incentive system (subsidies) that too frequently directs food production. He addressed the true cost of food effectively:

> Nobody wants food prices to go up, but if it is the case that the present low price of intensively produced food in developed countries is actually an illusion, only made possible by transferring the cost of cleaning up pollution or dealing with human health problems onto other agencies, then could correcting these anomalies result in a more beneficial arena where nobody is actually worse off in net terms? It would simply be a more honest form of accounting that may make it more desirable for producers to operate more sustainably,

particularly if subsidies were redirected to benefit sustainable systems of production.[144]

Prince Charles recommends "accounting for sustainability," which represents the true cost of food production; financial costs and the costs to natural capital – the earth's resources.

When our children discover industrial agriculture lacks the natural resources to produce food, they will ask the government for increased subsidies. Unfortunately, the government will be out of funds. What country would be willing to make loans that add to the immense U.S. debt? The U.S. is already a debtor nation; we just fail to act as one.

When our children discover their fields worn out, fresh water is unavailable, fuel costs are out of reach, fertilizer mines are exhausted and agricultural chemicals have ruined their waterways – will they agree that our fossil foods were cheap?

Biofuels

In the 1990s, the Clinton administration made a critical political mistake and stopped R&D on algae for food or biofuels. Those funds were shifted to corn ethanol for biofuel. The decision by the EPA to fund a corn ethanol industry may have been the most costly decision in American history because it accelerates natural resource depletion. When the U.S. runs out of resources to produce food, who will sell us food? Where will the government find the money to buy food for hungry Americans?

The farm lobby remains so strong that corn ethanol subsidies continue at around $20 billion a year, even though ethanol consumes more fossil energy than it returns. Huge subsidies flow primarily to large agribusiness and landowners, not to family farms. Subsidies continue in spite of clear scientific proof that corn ethanol is an expensive, wasteful proposition that not only massively depletes our natural resources but creates billions of dollars in degraded and damaged ecosystems.

The 44 million acres of corn grown for ethanol in 2010 could and should be replaced by less than 2 million acres of algae biofuel production, while improving air and water quality.

Climate Independent Foods

Biowar I: Why Battles over Food and Fuel Lead to World Hunger (Edwards, 2007) traces the money path, primarily to one company, ADM, that initiated the biofuel industry with millions in political donations to both parties. Today ADM receives billions each year in biofuel subsidies. A biowar occurs when a country burns food, typically as a horrific act of war on another country.

In Biowar I, the U.S. became the first country to burn its own food. Biowar I ignited when the Bush Administration announced the Energy Policy Initiative in 2005, which increased biofuel subsidies and mandates, Figure 10.2. The unintended consequence of producing large amounts of corn ethanol on U.S. and world food markets was predictably higher food prices. In the eyes of the UN, World Bank and most foreign countries, the U.S. ethanol policy contributed substantially to the terrible 2008 food riots in 40 countries.

Figure 10.2 Corn Burned for Ethanol and Food Stamps

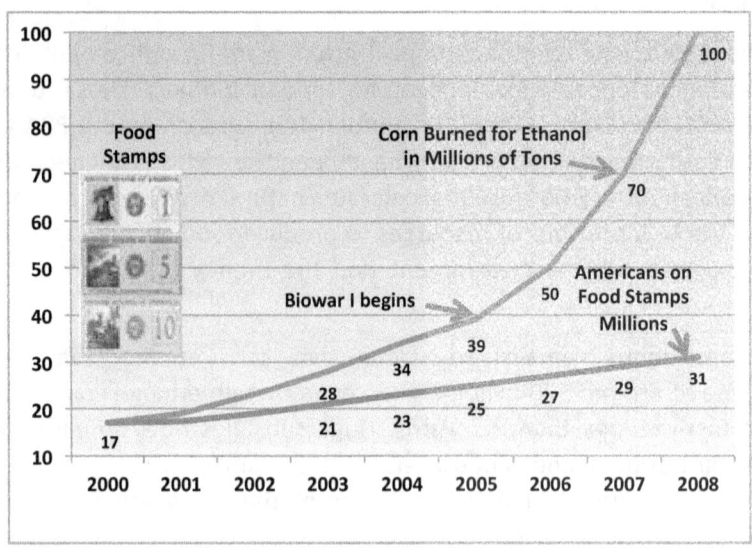

How could people in hungry countries not blame the U.S. for food shortages and price increases when prior to the ethanol program, America provided half the world's food grains and 70% of the world's corn imports? How does a country with over 60 million people receiving food support because they are hungry justify a policy of burning its citizen's food for a weak fuel additive? Over 43 million Americans are on food stamps and must abhor the concept of burning food because they know their $1 a meal buys less food.

In 2009, the U.S. became a net importer of food. A college sophomore could make the case that the U.S. biofuel policy is wasteful and foolish. We are burning our children's natural resources.

We may have developed a less expensive, food production system had people not had distaste to one word – algae.

Consumer behavior and algae

Why do most people have an immediate aversion to algae? The answer is false attribution. When asked to describe algae, people's top of mind typically elicits several words with strong negative connotations: "slimy, smelly, scummy and yucky." If putrefied raw meat were presented as steak, people would naturally dislike steak.

People falsely attribute the smell in ponds to algae because it certainly looks like algae. Actually, the odor comes not from algae but from the bacteria that attack and eat the algae. The bacteria consume all the oxygen algae added to the water, causing eutrophication. When the aquatic organisms are deprived of oxygen, they die and begin rotting, which adds smell and slime to ponds. Healthy algae give off lots of oxygen and smell similar to walking through a redwood forest – without the redwood trees, of course.

Weed algae in ponds grow in diverse communities of many microorganisms and are different from the algae we cultivate in abundance microfarms. Gardeners know they must remove the weeds from the garden or the weeds will take over the garden and consume all the nutrients. Abundance growers control weed algae in order to enable healthy production of the target species.

Consumers enjoy the taste of edible algae. At a recent Rotary meeting at a Kobe Steakhouse, the Rotarians were served three forms of algae at the luncheon: algae soup broth for taste and thickening and sea vegetables in the soup for texture, color and visual appeal. Separately a cold seaweed salad was served for color, taste and texture. The Rotarians were asked after lunch if anyone liked algae to eat, which

Climate Independent Foods

elicited the yuck factor, as all were extremely negative. They reversed judgment when they understood sea vegetables and algae salad provided the colors, taste and texture to their lunch. The Kobe manager brought out a large variety plate of sushi made with algae. The Rotarians consumed the sushi plate in one pass.

Nori, with a world market value of over $3 billion a year, represents only one of more than 1,000 sea vegetable. Nori serves as a luxury food. Cooks wrap Nori around a rice ball with seafood on the top. Toasting or baking brings out Nori's rich flavor and flakes complement rice or noodles. Epicurean cooks make a Nori soy sauce reduction for meat or seafood. Nori adds a spicy taste to jam and wine. Chinese cooks use Nori in soups and for seasoning fried foods. Many other sea and freshwater algae foods await commercialization.

Many forms of algae do not taste good by themselves. Some algae may need processing with other ingredients. Raw soybeans taste terrible but they are palate-pleasing when processed to tofu. Algae-based foods will become more attractive with informative food labels.

Food labels

Current food labels display only a few bits of information about calories and fat. We propose a more comprehensive food label that represents the nutritional value, sustainability, social value and ecological costs of food. Since the label is too long for many food products, we are working on a smart phone app that will allow consumers to access this label information for any food product.

We plan to expand the concept of taste tests to demonstrate the benefits of freedom foods. These blind taste tests will ask consumers first for their preferences based on standard consumer behavior protocols. The secondary test will ask for consumer preferences for food A or B using snack chip label similar to Table 10.1.

Table 9.1 Expanded Food Label
Example: 100 grams of Corn and Nori Chips

Food type	Industrial	Freedom
Nutrients	Corn chip	Algae chip
Protein	5 g	41.4 g
Calories	432 g	188 g
Calories from fat	16 g	3.7 g
Saturated fat	5.4 g	0.55 g
Polyunsaturated fat	3.9g	1.39 g
Monounsaturated fat	2.5g	0.20 g
Calcium	29 mg	280 mg
Magnesium	43 mg	300 mg
Vitamin A	10 mg	25 mg
Vitamin C	22 mg	210 mg
Niacin	5 mg	12 mg
Zinc	0.02 mg	5 mg
Dietary fiber	24 g	36 g
Health and nutrition		
Genetically engineered, GE	90% of U.S.	No
Health risk from GE	unknown	No
Nutrient dilution / empty calories	Yes	No
High in vitamins	No	Yes
High in minerals	No	Yes
High in antioxidants	No	Yes
High trace elements	No	Yes
Social justice		
Provides food security	No	Yes

Climate Independent Foods

Expanded food labels will provide consumers with additional information that is currently unavailable. Some consumers will ignore the expanded labels, just as they ignore the current brief labels. The purpose of the expanded label strategy is to support the dialogue on sustainable and affordable food and energy (SAFE) production.

Freedom foods are not a panacea and offer solutions to only some food issues. A set of needed technology breakthroughs, defined in the DOE National Algal Biofuels Technology Roadmap, will be required for the optimum use of algae biomass for commercial production.[145]

Chapter 11. eFootprint and Ecobalance Diet

How light is a butterfly's footprint?

The four primary metrics for sustainable food and biofuels are life cycle analysis, ecological footprint, freshwater footprint, and carbon footprint. LCA, also known as ecobalance, is a technique to assess environmental impacts associated with all the stages of a product's life from-cradle-to-grave. For food and biofuels, LCA examines environmental impacts from raw material extraction through cultivation, crop inputs, harvest, food transportation, refining, processing, supply chain, and disposal or recycling. Lifecycle analysis does not measure the threat of resource extinction and does not factor in financial, physical or weather risks.

An ecological footprint creates a metric for human demand on the Earth's ecosystems. The footprint compares human demand with our planet's ecological capacity to regenerate. For 2006, humanity's total ecological footprint was estimated at 1.4 planet Earths, 1.4 times renewal rate.[146] The full e-footprint calculation considers housing, transportation, recreation and food consumption.

Climate Independent Food

A freshwater footprint measures the consumptive use of freshwater. Recycled water for household use, (except yard and garden) that is reclaimed and reused is called non-consumptive water. Freshwater used for agricultural purposes and food processing is consumptive because the water is not available for reuse.

Water Footprint

A carbon footprint measures air pollution, as a derivative of the ecological footprint. A carbon footprint creates a metric for the total set of greenhouse gas emissions caused by any human activity. For foods, the carbon footprint typically expresses the amount of carbon dioxide, or its equivalent of other GHGs, emitted from production, transportation and consumption.[147]

The food e-footprint covers a wide variety of ecological factors but does not consider the possibility of resource extinction. Food production represents a special case were each of 24 fossil resources must be available to crops precisely on time or the crop fails.[148] When farmers find one fossil resource unavailable or unaffordable, they may lose their entire crop. Unfortunately, many of the vital fossil resources face extinction, especially in specific food growing regions.

The e-footprint for food consumption provides a broader metric than air pollution but narrower than the total ecological footprint. The e-footprint for food creates a metric that reflects the natural resources required to provide a consumer with food. The food e-footprint considers production, waste, risk, transportation, and pollution. The creation of a food e-footprint provides a means to measure one's impact on the planet. Most people are not aware of their food footprint. Self-awareness provides the first and necessary step for behavior change. Footprint calculations offer policy makers a set of standards that help in formulating sustainable food policy.

Food chain

People that eat high on the food chain consume large amounts of dairy and meat products and leave a large e-footprint. Vegetarians diminish their footprint far lower than meat consumers do, but still substantially higher than consumers that eat freedom foods.

Vegetarians leave a modest ecological footprint, not by choice but due to the way producers grow industrial foods. Each ton of grain consumes 1000 tons of freshwater, as well as considerable cropland, fuels, fertilizers and chemicals. Typically, about 60% of food is lost in the field or the food supply chain.[149] Fossil food production creates significant pollution, carries substantial physical and economic risk to farmers, continually extracts soil nutrients, erodes soil and pollutes ecosystems.

Fertilized soils release more than two billion tons of greenhouse gases every year, especially CO_2, methane and nitric oxide. Each cropland acre loses about 54 pounds of nitrogen, 13 pounds of phosphorus, 264 pounds of potassium and 132 pounds of calcium annually, which farmers typically replace with mined chemicals.[150]

Freedom foods avoid most of the resource consumption, pollution, risk, and weather problems that plague fossil foods. Growing freedom foods can repair air and water pollution and regenerate soils. Therefore, freedom foods consumers leave a tiny ecological footprint.

The peer-reviewed ecological footprint for food consumption remains to be constructed. The following draft model, Table 11.1, uses a hectare of corn and algae for comparison. Further refinement for the model may quantify the footprint per pound of protein. The food ecological footprint provides a relative rather than absolute metric for different consumers. This descriptive model serves as an educational tool because it captures the relevant categories that make up the footprint. Another model might use the food required to supply 2,200 calories per day.

Climate Independent Food

Table 11.1. Ecological Footprint for Food Consumption

Fossil inputs / hectare	Corn	Algae
Fertile soil – cropland	10,000 m², 2.47 acres	0
Fresh water	9.35 M lit, 2.47 M gal	0
Fossil fuels	7.6 liters, 2 gal	0.4 liters
Fossil fertilizer	91 kg, 200 lbs	0
Fossil chemicals	4.5 kg, 10 lbs	0
Transportation (ave.)	1,500 miles	50 miles
Shelf life	Days	Months
Loss to pests	15%	5%
Spoilage loss	40%	5%
Pollution		
Air – dust, CO_2, NO_x	Yes, tons	No
Poisons soil	Yes, massive	No
Poisons water	Yes, massive	No
Risk		
Crop failure	High	Low
30% yield loss	High	Low
Dust and pollens	High	Low
Poison exposure	High	Low
Hard physical labor	High	Low
Fatigue	High	Low
Heavy equipment	High	Low
Physical injury, disability or death	High	Low

Why not Freedom Foods Now?

Community health risk from pollution	High	Low
Soil		
Extracts nutrients	Yes	No
Extracts organics	Yes	No
Kills microorganisms	Yes	No
Breaks nutrient cycle	Yes	No
Degrades soil	Yes	No
Compacts soils	Yes	No
Erodes soil	Yes	No
Weather		
Yield loss; heat spike	Yes	No
Yield loss; cold spike	Yes	No
Drought venerable	Yes	No
Vulnerable to storms	Yes	Moderate

The USDA reports that in 2007, U.S. cattlemen used two billion bushels (112 billion pounds) of corn to produce 22.16 billion pounds of finished grain-fed beef. Farmers used 13.3 million acres to produce the feed grains, since corn production averages about 150 bushels per acre. Each pound of beef releases about 22 pounds of CO_2-equivalent greenhouse gasses.[151] Consequently, a single year of beef production releases roughly 2.5 trillion pounds of CO_2-equivalent greenhouse gasses. Cars add about 2.7 trillion pounds of new carbon the atmosphere each year.[152]

David Pimentel calculates a steer consumes about 100 pounds of grain per pound of edible beef produced.[153] Using the basic rule that it takes about 264 gallons of freshwater to produce one pound of hay and grain, about 26,400 gallons of freshwater are required to produce a pound of beef, Figure 11.1.

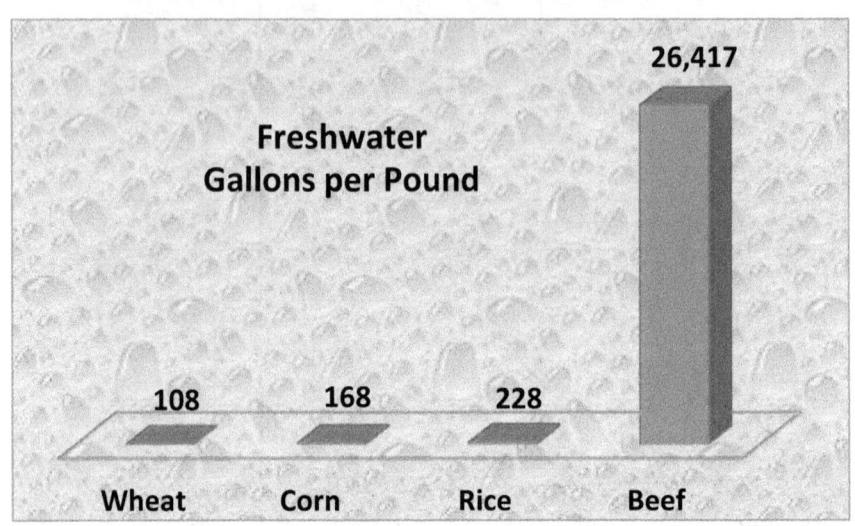

Figure 11.1 Freshwater cost of Food Grains and Beef

Biofuels also create a substantial environmental footprint. Biofuels compete with food, accelerate fossil resource depletion, and amplify soil erosion and pollution. Each gallon of ethanol consumes 3,000 gallons of water to produce the corn feedstock.[154] When drawn from fossil aquifers that do not replenish with annual rains, the water is not available for future generations – for food, biofuels or drinking. Much of the cropland west of the Mississippi River draws irrigation water from fossil aquifers and several will crash within a generation.

The Renewable Fuels Association reported that in 2010, producers converted 260 billion pounds of corn into 13 billion gallons of ethanol. Ethanol has 64% of the energy of gasoline so that is 8.3 equivalent gallons of gasoline. In 2010, U.S. farmers harvested nearly 400 million tons of grain, of which 126 million tons, primarily of corn, went to ethanol fuel distilleries. Each acre of corn production releases about 4500 pounds of CO_2. Therefore, ethanol feedstock adds another 700 billion pounds of CO_2 to the atmosphere. Of course, refining the corn feedstock to ethanol and burning the ethanol in vehicles adds additional CO_2. Ethanol production is neither green nor cleantech.

Another way to create the food ecological footprint uses a rating scale for each category. A peer-reviewed ecological footprint for food and biofuel production remains to be constructed. A recent *Applied Energy* article lays the groundwork for an e-footprint calculation.[155] The draft proposal here collapses several ecological categories to create a 100 point metric. Beef resides at the top of the food chain, because beef requires a high multiple of the resources required by grains and other foods low on the food chain, Table 11.2.

People eating high on the food chain consume large amounts of dairy and meat products leave a large e-footprint and earn a low ecobalance score. Freedom foods ecobalance diet scores are far lower than meat eaters since they are grown locally and avoid fossil resource consumption, pollution and moderate grower and weather risk. Freedom foods eaters earn high scores for foods that allow them to leave the e-footprint of a butterfly.

The ecobalance score for organic produce is probably around 40/100. Organic foods consume more fossil resources than industrial foods, except for fertilizers, chemicals and poisons. Some organic producers farm local to consumers and many enrich local communities.

The ecobalance score for corn used for food or biofuels approaches 70. Corn requires intense application of natural resources, is extremely pollutive. Corn requires ten times more nitrogen fertilizer than other food grains, which is not only consumptive but also extremely pollutive.

Algae-based biofuels could yield an ecobalance score 20 or lower if producers use waste or brine water for nutrients and waste CO_2 for their carbon source. Growers could also avoid most fossil fuels by using renewable energy.

Climate Independent Food

Table 11.2. Ecobalance Diet e-Footprint

10-point scale where 10 is low e-footprint consumption.	Beef Kilogram	Algae Kilogram
Fossil resource consumption		
1. Fertile soil – cropland	0	10
2. Fresh water	0	9
3. Fossil fuels	0	9
4. Fertilizers, chemicals and poisons	0	10
5. Transportation distance / spoilage	0	9
Pollution		
6. Air	0	10
7. Water	0	10
8. Soil	0	10
Risk		
9. Producers and community	0	9
10. Climate and weather tolerance	0	9
Total	**0**	**95**

Ecobalance scores do not account for the substantial food health issues or the use of GE crops. Michael Pollan, *The Omnivore's Dilemma*, Jeffrey Smith, *Seeds of Deception*, and Marie-Monique Robin, *The World According to Monsanto*, cover the health and GE crop issues effectively.

Chapter 12. Future Foods for any Climate

To cherish what remains of the Earth and to foster its renewal is our only legitimate hope of survival.
— Wendell Berry

Wendell Berry's plea to foster renewal for what remains of the Earth is critical for societies facing climate change, abandoned cropland, scarce water and polluted ecosystems. We stand at a crossroads in food production with two paths. We can:

1. Continue down the same path and produce primarily fossil foods relying on industrial agriculture, with full knowledge of the consequences for our children.

2. Develop new biotechnologies to grow freedom foods with abundance methods that preserve natural resources and clean polluted environments.

Fossil foods have considerable inertia. Farmers will continue using industrial methods to produce food – until the first of the 24 critical fossil resources becomes unaffordable or unavailable. Expanding human populations will need every conceivable food source to supply sufficient food for the 9.3 billion hungry mouths expected to share our limited resources by 2050.

Climate Independent Food

Climate chaos will create havoc on the food supply and push millions more farmers off their land. Farmers will deplete the modest fossil aquifer reserves and leave millions of cropland hectares abandoned with empty wells. Those once-fertile croplands will revert to dry prairie or desert. China, India and Africa provide living case studies for crashing aquifers and expanding deserts. The U.S. Mid-West and West will follow suit within the next generation.

People will have to migrate away from rural towns and cities when aquifers crash. Households need access to fresh water. Salt invasion from irrigation and sea surges will destroy additional cropland. No practical remediation exists for soil that becomes too saline to grow crops. Salt invasion, often accompanied by drought, has destroyed numerous prior civilizations.

Salt invasion ruins well water too. Tidal and storm surges often invade freshwater aquifers two or three times further than the surface water surge. Rising oceans will force millions from coastal cities and ruin prime coastal and delta croplands. Where will these people go?

Industrial agriculture's waste and pollution will continue to degrade and poison air, water and soil. Agricultural chemical and poison residue on foods will continue to cause children and adults to suffer illnesses and development disorders. Hidden hunger and empty calories will amplify the obesity and diabetes epidemic, creating severe drag for families, education and healthcare.

Novel solutions

Abundance methods that produce freedom food can avoid these predictable outcomes. Freedom foods are climate independent expand the discussion about sustainable and affordable food beyond organic production. Food policy strategy needs to include microcrops at the base of the food chain. Microcrops can be grown with plentiful resources that do not compete with industrial foods. Many microcrops are salt tolerant so they flourish in waste stream, brine or ocean water not useable by field crops.

Microcrops have been overlooked in the future of food debates because no one offered a practical plan to overcome the joint

Future Foods for any Climate

challenges of commercial production and consumer acceptance. A set of talking points in Appendix II summarizes the case for freedom foods. The cases presented here provide encouraging evidence but prove neither commercialization nor enthusiastic buyer behavior. The case for abundance methods and freedom foods has only just begun.

Abundance methods for food production have not yet benefited from government grants, government investment, Cooperative Extension Agent training or subsidies. If abundance received one tenth the financial and policy support currently given to industrial agriculture, freedom foods would be widely available within three years.

Too small?

Most farmers and farm policy leaders are unaware of freedom foods or abundance methods. As one USDA official said:

Microcrops are simply too small to make a difference.

We respectfully disagree. Farmers can domesticate and benefit from the millions of wild microcrop species that produce 40% of each day's new biomass. Each alga cell can produce over a million offspring a day. Growers can harvest half the biomass daily, all through the growing season in open microfarms or all year in closed systems.

These tiny plants produce many times more protein per year than land crops without using cropland or freshwater. Small is beautiful in nature as microcrops offer higher nutralence and are more easily digested than food grains. These tiny plants also deliver superior nutrition to animals and field crops. Hence, they reduce input costs for industrial farmers while improving food crop yields and quality.

Growers will be attracted to abundance for the larger yields of tastier and more nutritious produce for both algaculture and field crops. Gardeners and farmers are continually experimenting with new growing methods and plant varieties. Growers want to maximize productivity while practicing predominantly organic gardening. They also want to minimize input costs and the use of agricultural chemicals. Produce grown with abundance methods offers ecobalance and health differentiation superior to organic.

Consumers are curious about microcrops but have little opportunity to try them because there are no commercial producers yet. Consumers want to minimize their ecological footprint, possibly by following the ecobalance diet. Mothers want to feed their children foods that will improve their brains, eyes, reading skills and behavior.[156] Families would like to serve foods without poison residues that cause allergic reactions, dull brains and developmental disabilities. Families will prefer serving foods that fight obesity and diabetes instead of causing those and other "Western" diseases.

Older people prefer to choose foods that repair their brains and prevent cognitive decline and dementia.[157] These benefits accrue to consumers of freedom foods with high nutralence, including omega-3 fatty acids and other antioxidants.

Modern industrial foods have benefited from tens of billions of dollars in R3D, education and subsidies. Microcrops have received practically no government support and no education or subsidies. Algae production for biofuels has received less than $1 billion in government R&D funding but did not address food production.

NASA's $17.7 billion in annual funding explores space but produces no food. The current $300 million mission to Mars may discover evidence of ancient forms of life but no food. What is the logic for a country with one of five people on food support to fund space exploration when food R&D remains critical to the U.S. and global food security?

Critical tasks

Freedom foods will not magically appear because they appear to be a good idea. Serious work remains to convince growers to produce these crops and consumers to buy them. The first step is always the hardest – build new technologies that support distributed growers.

Abundance methods are technology neutral in the sense that effective production will require each grower's thoughtful application of farming methods that are appropriate locally. Farmers consider a broad set of environmental conditions as well as input availability and costs. Growers also monitor markets needs and buyer behavior.

Future Foods for any Climate

While abundance enables farmers to transition to predominantly organic methods, some may augment production with selected tools and techniques used in modern agriculture. For example, preliminary experience indicates growers may need to supplement one or more fertilizers because waste streams may not hold a full supply of replacement nutrients. Innovation research shows that incremental change occurs more successfully than a radical change like a swift move from industrial to organic farming.

Distributed small and medium size microfarm production offers the highest potential for local jobs and social justice. Large farms concentrate wealth in the hands of a few. Large agribusinesses limit innovation to the relatively few people engaged in the firm. Large firms create value for their stakeholders by hoarding their intellectual property, which serves the company but not society.

Farmers and gardeners are clever and innovative. They will adopt abundance methods when the production models are sufficiently developed and field-tested. Every grower wants freedom from weather, water and waste. Growers want to improve productivity, reduce costly fossil resources and eliminate waste and pollution.

R3D

Abundance needs R3D, research, development, demonstration and diffusion. Research and development must design and build not just the growing systems but even more critical, grower support systems. Demonstration sites must prove that microcrops consistently produce yields that are a multiple of field crops independent of weather or water. Growers must be able to produce superior food consistently with their local materials and inputs, in their climate and geography.

Peace microfarm diffusion to the many areas of need must wait until after the comprehensive research plan demonstrates growers have consistent success. Successful diffusion depends on positive grower experiences, and early adopter results will be critical.

Peace microfarms will employ a two-step flow strategy where Green Masterminds, initial growers, receive considerable training and support. Green Masterminds then will train their neighbors and community, which will multiply growers in every region. Peer training provides highly credible content for new growers. Remote monitoring will be critical to successful microfarm adoption.

Remote monitoring

R&D will beta test peace microfarms distributed in various locations. Growers will be supported with monitoring systems that use the Internet to send data to a lab staffed with algae experts, AlgaeCentral.org. The algae metrics system:

- Provides data on the culture health and vitality.
- Identifies when the culture needs addition inputs.
- Sounds an alarm if predators or weed algae threaten the culture.
- Guides growers with regular action recommendations.
- Provides the key metrics to validate productivity.

Microfarm monitoring and metrics will not only provide training and a safety net for growers but they will also provide transparency to those who may be skeptical of this new form of food production. Monitoring systems can build grower intelligence and a knowledge base that benefits all future growers. No practical knowledge base for non-scientists exists today for algae cultivation.

Algae metrics will monitor key culture parameters such as pH, temperature, culture density, nutrient availability and salinity. Growers will report harvest metrics in order to track production and microfarm productivity.

Algae metrics will assess the presence of grazers. Competing microorganism predators are normally present, but should remain sparse in healthy cultures. Green Masterminds will have an iPhone microscope picture app that allows them to take and send pictures of their culture in order to identify unusual microorganisms.

Weed algae also need to be monitored because, like terrestrial gardens, weeds compete for nutrients and diminish the percentage of the target compounds in the biomass. Weed algae are typically present but grower actions can keep the weeds from significantly diminishing production. In extreme contamination cases, the only

Future Foods for any Climate

defense for the grower is to harvest the algae, clean the microfarm, and start a fresh batch.

iPhone with attachable camera microscope

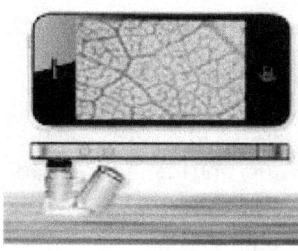 

Growers and research scientists will want to know a wide set of production parameters such as growth rates and harvest data. Each grower will need to monitor component analysis to understand what fraction of the biomass is lipids, protein, carbohydrate or other compound. Specialty producers, such as nutraceuticals will want to measure what fraction of the lipids are omega-3 or other antioxidants. Omega-3 producers may need further analysis to monitor the presence of each EPA, DHA and other target compounds.

Demonstration

Demonstration sites create the opportunity to share peace microfarm metrics in real time with environmental groups like the Sierra Club, Green Peace, World Wildlife Fund, The Heifer Project and the Environmental Working Group. Environmental groups need access to production metrics to validate the ability to grow clean food from sterilized waste streams. The United Nations and FAO need similar access to peace microfarm production data so report summaries can be integrated into food policy documents.

Environmental groups that focus on specific fossil resources can use peace microfarm production data to convey this new form of resource preservation. The Pacific Institute, led by Peter Gleick, publishes the annual *The World's Water*, but does not yet recognize abundance methods as a way to preserve scarce water resources.[158] The Sustainable Phosphorus Initiative, in the School of Sustainability at

Arizona State University promotes algae for nutrient recovery but has very little data on nutrient recovery or microfarms.[159]

Professional horticulture, agriculture, aquaponics and phycology organizations can inform their members about this alternative form of food production. Peace microfarms sited at strong agricultural research universities such as UC Davis, Cornell, Texas A&M and the University of Arizona can create strong production data for scientific publications. Universities and research institutes can analyze production data from distributed microfarms for ecology and economic comparisons with modern agriculture.

Siting peace microfarms in high traffic locations such as zoos, botanical gardens, science centers and aquariums will convey the microfarm value proposition to millions. ZooPoo can transform a huge cost, waste management, into a profit center for the zoo while creating a superb eco-exhibit. These sites can engage guests, students and volunteers in research projects as well as provide valuable production metrics. Public sites also offer extraordinary opportunity for targeted innovation on vital local issues such as sustainability, urban gardens and local jobs.

Special exhibitions, such as the Smithsonian and the Sustainable Food Exhibition in Milan, Italy in 2015 can demonstrate the value of peace microfarms to millions of visitors. The Algae Competition.com site offers excellent content for sustainable living and the future of food exhibits that may be located anywhere.

Microfarms can demonstrate urban abundance as growers produce food and feeds locally. Urban gardeners can recycle waste stream nutrients and clean water as they grow algae biofertilizers. Engaging Master Gardeners with smartcultures will validate increases in crop yield and quality while decreasing costs and waste.

Microfarms operating in government research locations such as the Cooperative Extension Service farms will demonstrate how these systems support field crops, reduce farming costs and minimize or eliminate waste streams. The Cooperative Extension Service can provide excellent test beds and innovation opportunities. Cooperative

Future Foods for any Climate

Extension agents understand horticulture and communicate with credibility to farmers.

Living Building Designed by Mark Buehrer, 2020 Engineering, features:

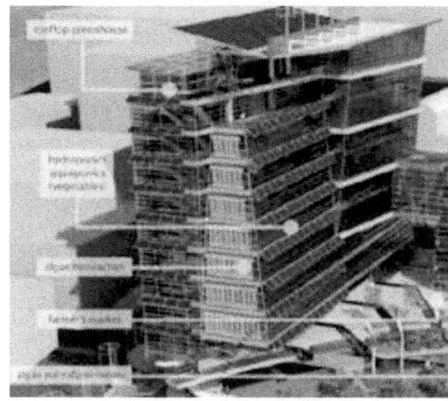

- Rooftop greenhouse
- Hydroponics for vegetables
- Algae microfarms
- Farmer's market
- Algae microfarm center
- Soil-based crops fed with algae biofertilizer

The EcoLab Algae Microfarm uses net-zero energy and produces net-zero waste.

Entrepreneurial opportunities

Entrepreneurs need to develop and demonstrate effective microfarms that allow diverse growers to produce good food and a multitude of coproducts. Microfarms need to be affordable growing systems usable by nearly anyone, practically anywhere. Growers have used open cultivation systems for decades, but not very efficiently.[160] Current growing systems are too expensive and require far too much expertise and labor to operate.

Peace microfarms will need comprehensive yet fast training for Green Masterminds. The training materials will include videos and tutorials demonstrating every step in the growing process to minimize the need for grower education and reading. Diffusion to the developing world requires that peace microfarms be very simple to operate.

Similarly, the monitoring and horticulture support system needs to be simple, yet assure consistent success. A strong support system will significantly reduce training time because Green Masterminds will receive advice at each stage of the growth cycle. Only a fast and easy to learn support system will motivate wide spread adoption.

Engineers know how to build microfarms but the capital cost, (Capex) needs to be reduced by a factor of ten. Current cultivation systems

Climate Independent Food

typically run above $10,000. Similarly, operational costs, (Opex), must be reduced and made easier, faster and better. Please engage in our fascinating global collaboratory, AlgaeFuture.org with your ideas for our abundant food future.

Quality control systems need to be transparent to insure that no producers in the distributed production network grow contaminated foods, feeds or other products. Extended food labeling will be important to assure consumers that the products produced are healthier for people, producers and our planet.

Policy leaders will need to navigate the complex government regulatory environments, including the FDA, USDA, OSHA and EPA. Industrial foods that are known to cause health problems are grandfathered in the food supply system. New foods like freedom foods that prevent health problems could be held up for years with very expensive bureaucratic regulatory hurtles.

Microfarmers will benefit from inclusion in new food legislation for small producers. Homemade Food Acts have been passed in 32 states and allow home-based entrepreneurs to sell up to $50,000 annually in "non-potentially hazardous" foods such as breads, jams and candy. Cottage food laws legalize practices such as sales at farmers' markets that are already commonplace in many communities but under the FDA radar. Cottage food acts enable entrepreneurs to test the market and then upgrade to commercial facilities and regulations when they expand production. Of course, these food acts will have to be amended to include algae foods.

Microfarm diffusion

Non-governmental organizations, professional groups and individuals will be helpful in diffusing microfarms to growers. Several groups including Antenna have built algae microfarms in Africa.[161] NGOs provide valuable help with funding, construction and training. The groups also give vital administrative and cultural support.

Service organizations including Rotary, Kiwanis and Engineers without Boarders have indicated a desire to assist with the diffusion of microfarms. Foundations such as Kellogg, Gates and Ford offer opportunities for both diffusion and applied research. Church service and development organizations may use microfarms to improve people's lives in countries where they have missions.

The Heifer Project may add microfarms to their list of transformative opportunities for needy families. Microfarms align with the Heifer Project mission to work with communities to end hunger and poverty and care for the Earth. Microfarms can provide high nutralence foods for people, their animals and their gardens.

In *Ending Hunger Now,* former Senator, Presidential candidate and U.N. hunger ambassador George McGovern suggests that the U.S. and other wealthy countries give more food aid to countries in need. He believes that the cost of hunger is unacceptable.

> *Today's malnourished pregnant and nursing mothers are producing tomorrow's barriers to personal, social and economic development – malnourished, brain dulled, listless children. Those fortunate enough to survive will dry through uncertain light, permanently diminished, unable to be productive, happy human beings."*[162]

Senator McGovern is absolutely right about the unacceptable cost of poverty but gifting food is unsustainable and creates dependence. Wealthy countries can and have gifted food in the short-term but soon will have insufficient food to give. No nation has the underlying fossil resources to sustain large food gifts. In the near future, no nation will have wealth to provide food transportation. The FAO reports that food transportation costs are often 400% higher than transferring the technology to grow food locally.

Rather than gifting food, why not send peace microfarms? Microfarms offer a far cheaper way to provide aid to hungry countries than transporting food. Peace microfarms may revolutionize foreign aid.

Disaster relief offers another excellent opportunity. When disaster strikes, local people are often left with bad weather, destroyed crops and infrastructure, lots of botanical waste and scarce freshwater. Microfarms can grow food quickly while remediating waste streams. In some disaster situations, the value of clean water may be higher than the food produced.

Climate Independent Food

Microfarms offer a novel way of providing food for food banks. Imagine a food bank growing half the bank's food locally. Schools and universities can use microfarms as learning and training tools while producing good food for students and faculty. Homeless shelters can employ people in microfarm production and produce food for their neighbors. Veteran's organizations may use microfarms for training and employment.

Prisons can change their waste stream cost to a profit center while training inmates for green jobs. America is the only country in the world with over twice as many prisoners as farmers. Prisons offer a great opportunity to educate green values while producing food. Special need farms have proven that food cultivation offers significant therapeutic value for a wide range of mental disorders commonly found among prisoners. The therapeutic value of freedom foods in prisons and special needs farms may exceed the economic value.

Sustainability goals

Freedom foods offer numerous sustainable outcomes besides independence from weather, water and waste. Possibly the most significant are health, social justice, new jobs and transportation fuel savings. Radically reducing obesity and diabetes may be a high priority for the many countries afflicted. Resolving obesity related diseases could save tens of billions in health care and educational costs and restore positive lifestyle for millions of families.

Freedom foods may not fully resolve some target diseases. However, current cognitive and medical approaches have not slowed the increase in diabetes or several other Western diseases. A novel healthy foods model without empty calories offers a new natural solution that warrants medical, scientific and public debate.

Transforming food production to a distributed model with many producers close to consumers will improve social justice and access to fresh good food. People who currently cannot grow their own food due to climate, available cropland or economics may have a stronger opportunity to produce for their family with microfarms.

Local production will create tens of thousands of new green jobs. Growers will produce fresh natural foods for their family, neighbors and farmers markets. High productivity will allow growers to produce large amounts of food in relatively small space. In addition, microfarms can be sited on land or buildings that have no or minimal alternative use, which reduces site costs.

Some green masterminds will sell their food products fresh, for direct consumption. Fresh foods may be made into dips or sauces such as aquamole, algaecream or algae ice cream. Others will sell alfu, which can substitute for tofu for use in soups, stews and salads. Algae flour products will include anything made with food grains, including bread, tortillas, chips, cakes and pies.

Aquamole producers and consumers in France

Food transportation cost reduction offers a valuable benefit set. The 50/50 freedom foods model with 50% of the food grown within 50 miles of consumers would save five times more liquid transportation fuel annually than the entire ethanol program. It would also save the 50 million acres of prime cropland planted in corn feedstock. Removing 50% of food trucks from our highways will substantially reduce pollution, road maintenance, and trucks that crush cars and people. A 50% reduction in black soot particulates in cities will reduce congestion, smog, and respiratory diseases significantly.

Local production reduces the 50 to 60% of food costs required in packaging, preservatives, storage, transport, shrinkage and spoilage. Consumers will benefit from healthier, fresh local foods. Unlike the medical solutions that require further R&D, the local freedom foods production model is not theoretical but practical. Local food production saves trillions of gallons of fuel, trillions of gallons of

Climate Independent Food

water, millions of cropland acres while preserving natural resources for our children.

The freedom foods sustainability goals will need public policy support in order to change from the tyranny of the present. Vested interests, big agribusiness, could take substantial losses in a new food production model. Only a few people monopolize farm policy today, while many people are victims of their reach for profits. Fortunately, this new food model will work best by engaging small, medium and large farms to work jointly for resource preservation.

Path forward

Our children's future depends on food security. Three critical threats to food security are global climate chaos, water scarcity and ecological pollution from agricultural waste. Our current food supply will be decimated by more fierce storms, hotter and more acidic oceans, rising sea levels, higher temperatures, hot dry winds and prolonged drought. The availability of affordable fresh water has already passed the tipping point in many good growing areas. Additional heat and drought will accelerate water loss and prove fatal for many farms, including in the U.S. Midwest and West.

Countries will go to war over the natural resources vital to food production unless farmers and food policy leaders find ways to preserve fossil resources. Access to sufficient water for food production threatens multiple countries. Developing a food supply independent of fresh water and that cleans polluted water makes strategic sense.

Many people cannot grow food today due to weather or the cost of crop inputs. Creating a climate independent food supply will allow growers to produce fresh food year round. Designing food production to recover and repurpose waste stream energy and nutrients will assure growers can afford crop inputs. Microfarms can moderate the waste and pollution caused by modern agriculture.

Green masterminds can grow high nutralence foods that end the nutrient deficiencies that severely limit growth and development of their children. Microfarms will help to clean ecosystems, which will improve the health and vitality of families that currently must live in waste stream plumes.

Farmers will embrace the opportunity to convert the costs associated with their waste stream to a profit center with the recovery and reuse of energy, nutrients and water. Society will celebrate the process of recovering energy and nutrients from waste streams while reversing air, water and soil pollution.

Urban growers will adopt abundance methods to produce freedom foods in their neighborhood. Developers will integrate microfarms in vertical farms, rooftop gardens and integrated living and growing spaces. Public architecture will become greenscapes with microfarms designed to provide visual, health and nutritional benefits.

Vertical farm   **Rooftop garden**

Public policy discussions will focus on net-zero use of fossil resources and net-zero waste. Microfarms that remediate waste streams will be designed into living buildings that serve multiple purposes while creating a net-zero ecological footprint. Consumers will look for fresh local foods with higher nutralence and no empty calories. Local markets will sell foods with superior nutrition and taste that are grown with no pollution or waste.

New industries will emerge to facilitate the engaging R3D that remain for peace microfarms. Entrepreneurs will develop new microfarm applications, new crops and new compounds such as medicines, nanomaterials and green chemicals. Green masterminds will find new pollution solutions that fit their communities. Growers will discover novel crops and diverse cultures that produce far more food than current methods. Smartcultures users will find more efficient ways for recovering farm or garden waste stream nutrients and delivering those nutrients to improve the yield and quality of field produce.

Freedom foods grown with abundance methods will not resolve all food issues. Unanticipated issues are certain to arise in abundance production, which is why monitoring systems are needed that are transparent and accessible. Microscope camera apps on mobile phones along with Internet connections make monitoring systems possible and practical.

Nature has used algae to support terrestrial plants for over 500 million years – since land plants evolved from algae. Now is the time that we need to mimic nature and enable algae to support sustainable, largely organic food production in a manner that uses fewer external inputs, lowers costs and reduces pollution.

Algae are available all over the Earth and are prepared to do their miraculous work to support growers, hungry consumers and society. We must act now to take advantage of algae's green promise – to create food security for everyone on our planet.

The design of affordable and easily operational microfarms remains our primary challenge. Collaborative open source social networks like www.AlgaeFuture.org can provide practical solutions. We can pool our knowledge and share it with those who will transform our food supply system – green masterminds.

Our collective actions will enable people globally to grow sustainable and affordable food and valuable coproducts for their family and community locally.

Appendix I Climate Change Impacts on Food

	Impacts on Food
Heat	Increased temperatures causes heat stress in food crops, which can significantly diminish their productivity and lead to plant death and crop failure.
Hot winds	Increased temperatures and dry winds evaporate soil moisture and increase the need for freshwater irrigation.
Water scarcity	Water, the critical resource for sustainable food production, has passed its tipping point as global warming causes food crops to need more water but water in many growing areas water sources have been degraded, depleted or diverted.
Rising sea levels	Oceans will consume millions of acres of prime cropland on coasts and river deltas and tidal and storm surges will destroy millions of acres of cropland from sea salt invasion.
Ocean acidity	Dissolved CO_2 in the oceans diminish fisheries, destroy shellfish and dissolve coral reefs that protect coasts and estuaries.
Higher ocean surface temperatures	Heat creates the energy that intensifies storms, hurricanes and typhoons. Heat also changes the rainfall patterns and leads to drought and severe forest fires experienced in the western U.S.
Extended spring and fall	Spring is starting a week earlier and fall lasts an extra week, enabling pest vectors – bugs, fungi, molds, mildews, viruses and weeds – to multiply earlier and sometimes survive the winter.

Climate Independent Food

Rain patterns	Shifts in rain patterns will cause huge losses of cropland that lack the infrastructure for irrigation.
Wildfires	Rangelands and forests are especially vulnerable to heat, drought and winds that drive catastrophic wildfires across the western U.S. states as well as Russia, China, India and Australia.
Loss of snow pack and glaciers	Snow packs are down 50%, which means faster run off and heavy flooding in the spring. Reservoirs, creeks and rivers may be only half full when irrigation is needed later in the growing season. Melting snow packs and glaciers mean less river water for irrigation and human use.
Blowing dust	While the U.S. Mid-West experienced severe flooding in 2008, Texas and Oklahoma lost millions of acres of crops to drought and blowing dust. Dust decimates crops, amplifies drought by removing soil moisture and erodes thin topsoil.

Appendix II. Freedom Food System Design Constraints
The Great Food Strategy Debate – Talking Points

Challenge	Description
Consumer Health	
1. Food security	Gives all people have access to affordable good food.
2. Consumer health	Delivers superior foods that naturally maximize health and vitality.
3. Fights diseases	Fights inflammatory diseases and improve major organ function including brain, eyes, heart, skin and respiratory system.
4. Avoids obesity	Low fat and cholesterol foods provide high quality protein, lipids, carbohydrates and other energetic compounds. Algae create satiety, which diminishes the urge to nosh.
5. Essential nutrients	Provides all 28 essential nutrients including essential vitamins, antioxidants, minerals and trace elements.
6. Nutralence	Delivers foods with high nutrient availability and density. Avoids hidden hunger and empty calories.
7. Sensory appeal	Foods have excellent color, aroma, texture, taste and mouth appeal.
8. Pesticide free	Clean and free of toxins, pathogens, fertilizers and poison residues.
9. Essential fatty acids	Micronutrients include omega-3 fatty acids and antioxidants.
10. Fresh and local	Over 50% of food can be grown within 50 miles of the consumer.
11. Transportation	Local production for 50% of food cuts

Climate Independent Food

	transportation costs and pollution.
12. Black soot	Truck traffic reduction reduces black soot.
13. Natural biodiversity	Natural biodiversity avoids GE seeds and genetic monocultures.
14. Carbon footprint	300% lower carbon footprint.
15. Ecological footprint	500% lower ecological footprint than industrial food.
Grower Health and Risk	
16. Grower health	Growers and their families are free from health risk from agricultural chemicals.
17. Grower safety	Growers are free from physical risk from heavy labor and industrial machinery.
18. Rural health	Growers, families, farm animals and neighbors are free from agricultural pollutants and poisons.
19. Grower risk	Growers and their families are relieved of production risk due to possible crop failure.
20. Dust and air pollution	Avoids dust and air pollution by not disturbing the soil.
21. Waste	Avoids nutrient and water waste by recycling the culture. The only thing given off by algae production is pure oxygen.
22. Productivity multiple	Grower consistently produces 20 to 30 times more food per unit of cultivation every year independent of weather.
23. Affordable costs	Input costs are affordable and not tied to the price of fossil resources like fuel.
24. Nutrient extraction	No nutrient extraction because crops are grown in water on non-croplands.
25. Sustainable	Freedom foods offer sustainable

Future Foods for any Climate

production	production for at least 7 generations.
Ecological health	
26. Fertile soil	Grows in containers that can be placed on non-cropland.
27. Fresh water	Grows in brackish, brine, waste and ocean water and can avoid freshwater.
28. Fossil fuels	Uses little energy that can be delivered from renewable sources.
29. Fertilizers	Can recover nutrients from waste streams and avoid mined inorganic fertilizers.
30. Poisons	Grow without pesticides, herbicides and fungicides by using natural biopesticides and controlling culture parameters.
31. Air remediation	Cleans air of gasses and black soot.
32. Cleans water	Cleans polluted water.
33. Soil regeneration	Restores soil fertility.
Sociological health	
34. Social equity	Everyone has access to affordable good food or the inputs to produce food.
35. Nutrient deficiencies	FAO reports that 50% of children globally suffer from nutrient deficiencies. Freedom foods can end nutrient deficiencies.
36. Economic risk	Growers know their crops will produce and not fail. Crop inputs are affordable.
37. Political risk	Governments and corporations do not control of food distribution and access.
38. Local jobs	Distributed peace microfarms can grow food nearly anywhere; growers enjoy good jobs.
39. Community	Growers can form cooperatives to produce

health	vitamins, minerals, nutraceuticals, vaccines, pharmaceuticals and medicines to promote family and community health.
40. Food independence	Distribution of freedom food knowledge globally will enable people to grow good food and other products for their family and community locally.
	Smartcultures Benefits – Algae Biofertilizer for field crops
42. Yield	Increase yield 20 to 80%.
43. Size	Increase produce size 20%.
44. Germination rate	Increase germination rate 20%.
45. Growth rate	Increase growth rate 10%.
44. Maturity	Speed time to grow to maturity 10%.
45. Color	Improve color 20%.
46. Sensory value	Improve aroma, taste and texture 20%.
47. Self life	Increase self-life 30%.
48. Water consumption	Smartcultures cut field crop water consumption by 20 to 30%.
49. Fertilizer consumption	Reduce fertilizer requirement by 60% for N and 40%+ for P and K. Micronutrient costs go to zero through waste stream recovery.
50. Pesticide use	Smartcultures can decrease the need for pesticides by 50% and fungicides by 75%.
51. High pH	Algae biomass that grows in the field can reduce elevated soil pH to normal levels.
52. Compression	Smartcultures deliver algae biofertilizers that can increase soil porosity by 500%.
53. Salt invasion	Improves soil porosity and allows irrigation

	salts to percolate below the root zone.
54. Humus loss	Rich algae biomass grows in the field, adding organic matter and natural fertility.
55. Pest control	Biofertilizers and growth hormones enable plants to produce their own biopesticides.
56. Crop stress	Algae biofertilizers improve plant health and vitality. Crops are able to withstand weather stress as well as pest vectors.
57. Erosion	No erosion from wind or water because the soil is not disrupted or degraded.
58. Pollution	Everything is recycled and repurposed, leaving no waste to pollute.

Acknowledgements

New ideas build on the considerable research provided by science and environmental prior pioneers, including:
- David and Marcia Pimentel, emeritus, Cornell University
- Lester Brown, President of the Earth Policy Institute
- Jeffery Sachs of the Earth Institute at Columbia University
- Fred Krupp, president, the Environmental Defense Fund
- Ken Cook, President of the Environmental Working Group

Professors Qiang Hu, Milton Sommerfeld and Bruce Rittman from Arizona State University supported questions on molecular biology and algae production.

Science	Business – Econ.	Agribusiness
• Al Darzins	• Dan Simon	• Jon Ewen
• Mike Siebert	• David Schwartz	• Richard Morrison
• Jim Lane	• Mike Pasqualetti	• Ben Cloud
• Dan Childers	• Mark Allen	• Gary Wood
• James Elser	• Alan Resnik	• Doug Young
• Carol Johnston	• Gary Dyer	• Jim Robertson
• Chris Low	• Herb Roskind	• Tracy Penwell
• Andy Ayers	• William Cockayne	• Barry Spiker

Also helpful were the published works of Paul Ehrlich, Sandra Postel, Nobel Laureate Al Gore, Harvey Blatt, Fred Pearce, Michael Pollen, Brian Halweil, Clay Jason and Linda Graham. High-content websites were a great support such as Algaebase, U.N., W.H.O., the National Resources Defense Council, Sierra Club, Green Peace, Audubon Society, Union of Concerned Scientists, Center for Energy and Climate Solutions, Clean Water Network and Public Citizen. Also useful were government sources including DOE, EPA, U.S. DA, NOAA and NREL.

Mark Edwards

Mark designs nutritious, sustainable and affordable food and energy (SAFE) production systems available to all people on Earth. He pursues abundance; to create food security for all and to help growers leave every field better than they found it.

Mark graduated from the U.S. Naval Academy where he earned degrees in engineering, oceanography and meteorology and Jacques Cousteau motivated and mentored his interest in the oceans and global stewardship. He holds an MBA and PhD in marketing and consumer behavior and has taught food marketing, engineering, sustainability, agribusiness, leadership and entrepreneurship at Arizona State University for over 30 years.

Mark served as marketing director for the Longevity Research Institute directed by Nathan Pritikin. The LRI focused on actions designed to improve the diet and exercise behaviors for people with health needs. The work led to the Pritikin diet and to Pritikin health foods. He also served as a director for a Fortune 50 foods company and has done extensive R&D on new foods, sources and consumer behavior. He has consulted for Monsanto, DuPont, Nabisco, Quaker Oats, General Mills, Borden, Coca-Cola, Frito-Lay, Disney, GE, Intel, J&J, Merck, GM, Bank of America, and many other companies.

He has published over 120 articles and 21 books that span business and science disciplines. His 360° *Feedback* was a business best seller. Several have won best science and environment book awards including *Green Algae Strategy, Abundance: Sustainable fossil-free Food, The Tiny Plant that saved our Planet* and *Freedom Foods*. Colleges in in over 30 countries use some of the Green Algae Strategy series in food, energy and sustainability courses. Mark cofounded the global AlgaeCompetition with Robert Henrikson to enable people globally to produce food and coproducts for their family and community locally.

Robert Henrikson

Robert Henrikson has been a green business entrepreneur for over 30 years in sustainable development business models for algae, bamboo and natural resources. Robert was a founder of Earthrise Farms and for 20 years, was President of Earthrise, a pioneer in algae.

Robert was the creator and director of the International Bamboo Building Design Competition (BambooCompetition.com), and the former CEO of a leading company building certified, code-approved bamboo buildings. Robert is the co-author of the book *Bamboo Architecture in Competition and Exhibition* based on the International Bamboo Building Design Competition.

Robert serves as an AlgaeAlliance.com consultant on business development, strategic planning, branding, sales and marketing, advising companies and investors in algae ventures. He developed Earthrise® brand products in the USA and 30 countries. Authored the book *Spirulina World Food* in 2010, previously *Earth Food Spirulina*, translated into 6 international editions (SpirulinaSource.com).

Robert has written numerous articles and produced many videos on algae over the past 30 years, and currently contributes articles to Algae Industry Magazine and speaks at algae conferences. In 2011 he launched the International Algae Competition: A Global Challenge to Design Visionary Algae Food and Energy Systems.

Robert is a photographer (Panmagic.com) and documentary filmmaker, and produced the DVD series *Folding Time and Space at Burning Man* (Folding-Time.com). He is co-owner of Hana Gardenland, a botanical paradise retreat in Hana Maui, with vacation retreats and eco-tourism (HanaPalmsRetreat.com and Wild Thyme Farm, a sustainable forestry and permaculture farming eco-community. Email: roberthe@sonic.net.

Future Foods for any Climate

The Green Algae Strategy Series

Mark R. Edwards

The **Green Algae Strategy Series** focuses on creating Sustainable and Affordable Food and Energy – "SAFE" production. **The Green Algae Strategy Series** are available for free downloading in color PDF for students, teachers and food and energy policy leaders at www.algaefuture.org. They are also available on Amazon.com and other retailers. Teachers, professors and policy leaders use these SAFE production books in schools and colleges globally for courses in sustainability, engineering, business, politics, social entrepreneurship, food, water, energy, ecology, environment and world future.

BioWar I: Why Battles Over Food and Fuel Lead to World Hunger, 2007. BioWar I, where food is burned for fuel, must be ended by withdrawal – not of soldiers, but of damaging agricultural subsidies.

Green Algae Strategy: Engineer Sustainable Food and Fuel. 2008. Algae offer solutions for sustainable and affordable food and energy because algae are the most productive biomass source on Earth. **Best Science Book – 2009, Independent Publisher Awards.**

Green Solar Gardens: Algae's Promise to End Hunger, 2009. Algaculture in small but beautiful solar gardens and algae microfarms distributed globally will enable SAFE production locally.

Crash: The Demise of Fossil Foods and the Rise of Abundance. 2010. Traditional fossil-based agriculture sits precariously on a foundation of unsustainable fossil resources that will become unaffordable and then will run out. Abundant agriculture is sustainable because it uses plentiful inputs that are cheap and will not run out.

Smartcultures: Nature's tiny Genius – Algae – Reverses Pollution and Regenerates Degraded Ecosystems, 2011. Farmers may recycle farm wastes to their fields using abundance microfarms. Smartcultures give 20 – 30% higher yields by providing bioavailable nutrients at just the right time. Farmers save 30 – 40% by reducing input costs and reduce ecological pollution by 90%.

Climate Independent Food

Abundance: Sustainable Fossil-free Foods with superior Nutrition and Taste; less Pollution and Waste, 2010. Abundance presents the value proposition for algae food, feeds and other forms of energy using plentiful resources that will not run out. Abundance growers can clean the air and water while they grow foods with superior nutrient density and better sensory values, including color, texture and taste. **Pinnacle Gold Medal winner 2011 for the best Environmental book.**

The tiny Plant that saved our Planet. The incredible true story of Tiny, Mighty Al. Tiny Mighty Al saved our planet by eating the bad carbon genie, which enabled the earth to cool and gave us oxygen. Al saved us again by becoming the bottom of the food chain and providing all living creatures with nutritious food. If we educate our children, maybe they will prompt us to take action — now. **Nautilus Silver Medal winner 2011 for the best children's book.**

Abundant Agriculture: Smartcultures enable superior Nutrition and Yields from Regenerated Fields, 2010. Abundant agriculture represents the first new form of agriculture in 60 years. Abundant agriculture produces sustainable food, feed, fiber and other coproducts using primarily non-fossil resources that are plentiful, affordable and often surplus. Abundant agriculture growers use abundance methods to produce healthy, nutritional foods.

Freedom Foods: Superior Nutrition and Taste from low on the Food Chain for People, Producers and Our Planet, 2011. Freedom foods liberate consumers to make healthier food choices. Freedom foods are sustainable and grow free of fossil resources, GMO material and agricultural chemicals and pesticides.

***Imagine Our Algae Future:* Visionary Algae Architecture and Landscapes**, 2012, with Robert Henrikson. See visionary images from the AlgaeCompetition.com showing how algae will change our world. Contestants from 40 countries created amazing graphics, pictures and videos showing how algae is produced today and will be used tomorrow for food, feed, biofuels, medicines and ecological repair.

References

[1] Edwards, Mark R. *Abundance: Sustainable Fossil-free Foods,* 2010.
[2] Manning, Richard. *Against the grain: how agriculture has hijacked civilization,* New York, North Point Press, 38.
[3] Pimentel, David and Marsha Pimentel. *Food, energy and society,* third edition, New York: CRC Press, 2008, 27.
[4] USDA, Food stamps make America stronger, 2010, www.fns.usda.gov/
[5] Morgan Stanley, Little minds need big meals, *Wall Street Journal,* Dec 15, 20011, A1.
[6] Feed America, http://feedingamerica.org/SiteFiles/child-economy/
[7] Scherr, Sara and Sajal Sthapit. Farming and land use to cool the planet, in The State of the World 2009, the World Watch Institute, 2009, 37.
[8] Kolbert, Elizabeth. *Field Notes from a Catastrophe: Man Nature and Climate Change.* New York: Bloomsbury, 2006: 96-97.
[9] Diamond, Jared. Collapse: *How Societies Choose to Fail or Succeed.* New York: Viking Adult, 2004, 54.
[10] IPCC 4th, Working Group I: The Physical Basis for Climate Change, Summary for Policy Makers, 2010.
[11] Anderegg, WRL, et al. Expert credibility in climate change. *Proceedings of the National Academy of Sciences U.S.A.* 107, 2010, 12107-12109
[12] Mohan K. Wali et al., "Assessing Terrestrial Ecosystem Sustainability," Nature & Resources, October–December 1999, pp. 21–33.
[13] Lobel, David and Gregory Asner. "Climate and Management Contribution to Recent Trends in US Ag Yields." *Science* 299 (14 Feb. 2003): 1032.
[14] J. Larsen, Earth Policy Institute, published online 28 July 2006 www.earth-policy.org/Updates/2006/Update56.htm
[15] W. Easterling et al., in Climate Change 2007: Impacts, Adaptation, and Vulnerability, M. Parry et al., Eds. (Cambridge Univ. Press, NY, 2007), 976.
[16] Tankersley, Jim. California farms, vineyards in peril from warming, U.S. energy secretary warns, Los Angeles Times, February 4, 2009.
[17] Scherr, Sara Farming and land use to cool the planet, 31.
[18] World Bank, Agriculture Development, World Bank, Washington, DC, 2008.
[19] Battisti, David and Rosamond Naylor. Historical Warnings of Future Food Insecurity, *Science,* 8 September 2008,10.1126
[20] http://www-ce.ccny.cuny.edu/nir/sw/hardiness-change.html
[21] Gleick, P. *The World's Water 2001.* Washington, DC: Island Press, 2000: 52.
[22] Ibid., 52.

[23] Garrett Hardin. The Tragedy of the Commons. *Science,* 1968, 162 (3859): 1243–1248.
[24] California Dept. of Water Resources, 2009. ww.owue.water.ca.gov/agdev/
[25] Gorman, Steve. California farms lose main water source to drought, Reuters, February 20, 2009.
[26] Burke, Garance. Will drilling more wells in California help or hurt? The Associated Press, January 11, 2010.
[27] Ten, Kate, Daniel. Grain prices soar globally, *The Christian Science Monitor.* March 27, 2008, 1.
[28] FAO, ResourceSTAT, electronic database, at faostat.fao.org/site/405/default.aspx, updated 30 June 2007.
[29] Sharma, M. C. and Owen, L. A., J. Quat. Sci. Rev., 1996, 15, 335–365.
[30] MacKenzie, Debora. Melting glaciers will trigger food shortages, New Scientist, 20 March 2008.
[31] Cole C.V, et al. Global estimates of potential mitigation of greenhouse gas emissions by agriculture. Nutr. Cycl. Agroecosyst. 1997;49:221–228. doi:10.1023/A:1009731711346
[32] USDA, ERS, http://www.ers.usda.gov/Data/FertilizerUse/
[33] Uri, N. D. "Agriculture and the Environment – the Problem of Soil Erosion," *Journal of Sustainable Agriculture,* 16;4, 71-91.
[34] FAO, Meat and meat products, *Food Outlook* FAO, Rome, December, 2006.
[35] Steinfeld, H. and Shalonda, P. Old players new players, *Livestock report 2006*, Rome: FAO, 2006, 3.
[36] Steinfeld, H. *Livestock's long shadow: environmental issues and options*, Rome: FAO, 2006.
[37] U.S. Environmental Protection Agency. *Rivers and Streams,* in National Water Quality Inventory: Report 2000. U.S. EPA. August 2002: 13-14.
[38] Thelen, Kurt D. *What is the Direct Carbon Footprint of Biofuel Relative to Gasoline?* Michigan State University, Crop & Soil Sciences, https://www.msu.edu/~thelenk3/Acrobat/C%20footprint.pdf
[39] Pegg, J.R. *U.S. Corn Production Feeds Expanding Gulf Dead Zone,* ENS, http://www.ens-newswire.com/ens/jun2008/2008-06-18-092.asp
[40] DNR, *State of Iowa, Public Drinking Water Program 2006 Annual Compliance Report,* (Des Moines, IA: June 2007.
[41] Sutter, John David. Five Water Bodies not Polluted, *News Oklahoma.com,* June 8, 2008, 1.
[42] Jamal, G. A., et al. Low level exposures to organophosphorus esters may cause neurotoxicity. *Toxicology* 181-182: 2002, 23–33.

[43] Prince Charles, Future of Food Conference, Georgetown University May 4, 2011. http://washingtonpostlive.com/conferences/food
[44] Ibid.
[45] Lawrence, Felicity. Not on the Label Penguin. 2004, 213
[46] Hidden Hunger, http://www.micronutrient.org/english/View.asp?x=573
[47] Engel SM, et al. 2011. Exposure to Organophosphates, and Cognitive Development in Childhood. Envir. Health Persp. 10.1289/ehp.1003183.
[48] Brennan, Morgan. America's Most Polluted Cities, Forbes, 04.28.11.
[49] Trasande, Leonardo and Yinghua Liu. Reducing the Staggering Costs Of Environmental Disease In Children, Estimated At $76.6 Billion In 2008, *Health Affairs*, April 2011.
[50] GE crops and pesticide use, http://www.ucsusa.org/food_and_agriculture/science_and_impacts/
[51] http://news.uns.purdue.edu/html4ever/0012.Huber.deficiency.html
[52] Pimentel, David. Food, energy and society, 75.
[53] Young, A. Agroforestry for soil conservation. Wallingford, UK: CAB, 1989.
[54] Wilkinson, Bruce H. and Brandon J. McElroy, The impact of humans on continental erosion and sedimentation, Geological Society of America Bulletin, January 2007; v. 119; no. 1-2; p. 140-156; DOI: 10.1130/B25899.1
[55] Natural Resources Conservation Service, Soil Erosion, 2006.
[56] EPA, Assessed Waters, in 2000 National Water Quality Inventory, EPA-841-R-02-001, EPA, Office of Water, Washington, D.C., August 2002.
[57] Zwerdling, Daniel. India's Farming 'Revolution' Heading For Collapse, NPR, April 13, 2009. http://www.npr.org/templates/story/story.php?storyId=102893816
[58] Sainath, P. The Largest Wave of Suicides in History, Counterpunch, February, 12, 2009. http://www.counterpunch.org/sainath02122009.html
[59] World Hunger Statistics 2010, http://www.worldhunger.org
[60] http://www.bread.org/hunger/global/
[61] Young Richard, A DeVoe Jennifer E. Who Will Have Health Insurance in the Future? An Updated Projection. *Annals of Family Medicine*, 10;2, 156-162, April 2012.
[62] CDC data and statistics on obesity and diabetes in America. http://www.cdc.gov/features/dsObesityDiabetes/
[63] http://www.cdc.gov/diabetes/pubs/pdf/ndfs_2011.pdf
[64] Environmental Working Group, Crop Subsidies Data Base, http://farm.ewg.org/
[65] Cook, Ken. Government's continued bailout of corporate agriculture, http://farm.ewg.org/summary.php

[66] Food Planet, Food summit blames trade barriers, biofuels, June 4, 2008, www.planetark.com/dailynewsstory.cfm/newsid/48626/story.htm

[67] Edwards, Mark R. *BioWar I, Why Battles Over Food and Fuel Lead to World Hunger,* CreateSpace, 187.

[68] Joffe-Walt, Chana. Why U.S. subsidies Brazil's cotton crop, Nov 9, 2010. http://www.npr.org/blogs/money/2011/01/26/131192182/cotton

[69] Edwards, Mark R. *Abundance: Sustainable Fossil-free Foods,* 2010.

[70] http://www.agcensus.usda.gov/Publications/1997/Agricultural_Economics_and_Land_Ownership/indexintro.asp

[71] OECD Highlights Chinese Pollution. *Financial Times,* 17 July 2007. http://www.ft.com/cms/s/932c36ca-348c-11dc-8c78-0000779fd2ac.html.

[72] Conkin, Paul K. *A Revolution down on the farm: The transformation of American Agriculture since 1929,* U. of Kentucky Press, 2008, 164.

[73] Edwards, Mark, Algae101 Blog, Algae Industry Magazine, www.algaeindustrymagazine.com/algae-101-part-27-soar/

[74] Pimentel, D., Hepperly, P., Hanson, J., Douds, D., and R. Seidel. 2005. Environmental, energetic, and economic comparisons of Organic and Conventional farming systems. *Bioscience,* 55(7):573-582.

[75] Resources for the Future, *What Do the Damages Caused by U.S. Air Pollution Cost?* December 17, 2007, http://www.rff.org

[76] Neuman, William. High Prices Sow Seeds of Erosion, *New York Times,* April 12, 2011.

[77] Iowa Daily Erosion Project, 2011, http://wepp.mesonet.agron.iastate.edu/index.phtml

[78] Edwards, Mark R. *Abundant Agriculture: Smartcultures enable superior Nutrition and Yields from Regenerated Fields* Tempe: CreateSpace, 2010.

[79] Goyal, SK. A profile of algal biofertilizer. in *Biotechnology of Biofertilizers,* Kannaiyan, S. Ed., Delhi: Narosa Publishing House, 2002, 250 – 258.

[80] Edwards, Mark R. *Green Algae Strategy: Engineer Sustainable Food and Fuel,* CreateSpace, 2008, 44.

[81] A list of algae collections is available at AlgaeCompetition.com.

[82] Edwards, Mark. *Green Solar Garden: Algae's Promise to End Hunger,* 2009.

[83] Edwards, Mark. *Abundance,* 2010.

[84] Edwards, Mark. *Freedom Foods: Superior Nutrition and Taste from low on the Food Chain for People, Producers and Our Planet,* 2011.

[85] Warner, Jennifer. CDC: Kids Lack Access to Healthy Food Choices, WebMD Health News , April 26, 2011.

[86] Center for Food Safety, http://www.centerforfoodsafety.org /2011/03/18/

[87] http://www.cdc.gov/diabetes/pubs/pdf/ndfs_2011.pdf

[88] Pimentel, David and Marsha Pimentel. *Food, energy and society*, 3rd edition, New York: CRC Press, 2008, 27.
[89] http://www.ers.usda.gov/briefing/organic/Farmsector.htm
[90] Edwards, Mark R. *Abundant Agriculture*, 2010.
[91] Cui-Hua Qi, Min Chen, Jie, Soong, Bao-Shan Wang. Increase in aquaporin activity is involved in leaf succulence of the euhalophyte suaeda salsa, under salinity. *Plant Science*; 176;2, pp. 200-205, Feb 2009.
[92] Nutralence is a new word that refers to plants that naturally store nutrients, delivering higher availability and nutrient density.
[93] Jeff Norrie, Seaweed Research. *American Fruit Grower*, 128;3, pp. 48-50, Mar 2008.
[94] Luescher-mattli, M. Current Medical Chemistry-Anti-Inflammatory Agents, in Ingenta-Connect, 2, 2003, 219-225.
[95] MacArtain P, Gill CIR, Brooks M, Campbell R, Rowland IR. Nutritional value of edible seaweeds. *Nutrition Review*, 2007, 65:535-543.
[96] Spolaore P, Joannis-Cassan C, Duran E, Isambert A. Commercial applications of microalgae. 2006 *J Biosci Bioeng* 101:87-96.
[97] Garcia-Casal, et. al. 2007.
[98] Garcia-Casal MN, Pereira AC, Leets I, Ramirez J, Quiroga MF. High iron content and bioavailability in humans from four species of marine algae. *Journal of Nutrition,* 2007, 137:2691-2695.
[99] Kavaler, Lucy. *Green Magic: Algae Rediscovered*. Thomas Crowell, NY, 1983, 99-101.
[100] Dhargalkar, V. K. and X. N. Verlecar. Southern Ocean seaweeds: A resource for exploration in food and drugs. *Aquaculture*, 287(3/4): 2009, 229-242.
[101] Lisheng, L. et. al. Inhibitive effect and mechanism of polysaccharide of spirulina on transplanted tumor cells in mice. *Marine Sciences*, Qindao, China. N.5, 1991, 33-38.
[102] Luescher-mattli, M., Current Medical Chemistry-Anti-Inflammatory Agents, 2010, 2, 219-225.
[103] Palmquist RE. 2008. Apparent response to homotoxicology, salmon oil and blue-green algae in a single geriatric canine case of episodic mentation changes. JAHVMA. 27 (1): April-June, 2010, 10-15.
[104] McCarty MF, Barroso-Aranda J, Contreras F. NADPH oxidase mediates glucolipotoxicity-induced beta cell dysfunction--clinical implications. Med Hypotheses.;74(3):March, 2010, 596-600.

[105] Lee EH, Park JE, Choi YJ, Huh KB, Kim WY. A randomized study to establish the effects of spirulina in type 2 diabetes mellitus patients. Nutr Res Pract. 2008 Winter; 2(4): 295-300.

[106] Muthuraman P, Senthilkumar R, Srikumar K. Alterations in beta-islets of Langerhans in alloxan-induced diabetic rats by marine Spirulina platensis. J *Enzyme Inhib Med Chem.* 2009 Dec;24(6):1253-6.

[107] Garbuzova-Davis, Svitlana. *The Open Tissue Engineering and Regenerative Medicine Journal*, 2011, (3:36-41).

[108] Gupta S, Hrishikeshvan HJ, Sehajpal PK. Spirulina protects against rosiglitazone induced osteoporosis in insulin resistance rats. Diabetes Res *Clin Pract.* 2010 Jan;87(1):38-43.

[109] Shytle DR, Tan J, et al. Effects of blue-green algae extracts on the human adult stem cells in vitro. *Med Sci Monit.* 2010 Jan, 16(1): BR1-5.

[110] Edwards, Mark R. *Abundance,* 54.

[111] Fountain, Henry. Too Many Small Fish Are Caught, Report Says, *New York Times*, April 2, 2012.

[112] http://www.oceanconservationscience.org/foragefish/

[113] http://www.dietresearch.com/

[114] Edwards, Mark R. *Abundance*, 91.

[115] Milton K. The critical role played by animal source foods in human (*Homo*) evolution. *J. Nutr.* 2003, 133:3886–92S.

[116] Edwards, Mark R. Algae 101 Blog, Part 26: Did Algae's Great Taste Make us do it? *Algae Industry Magazine*, May, 2011.

[117] Kulshreshtha, Archana et. al. Spiralina in healthcare management, *Current pharmaceutical biotechnology*, 2008, nine, 400 – 405.

[118] MacArtain P, Gill CIR, Brooks M, Campbell R, Rowland IR. Nutritional value of edible seaweeds. Nutr Rev 2007, 65:535-543.

[119] Yamada Y, Miyoshi T, Tanada S, Imaki M. Digestibility and energy availability of Wakame seaweed, *Jap J Hygiene*, 2007, 46;788-793.

[120] Edwards, Mark R. *Abundance*, 88.

[121] Yangthong M, et al. Antioxidant activities of four edible seaweeds from the southern coast of Thailand. *Plant Foods Hum Nutr* 2009, 64:218-223.

[122] Abad MJ, Bedoya LM, Bermejo P. 2008 Natural marine anti-inflammatory products. Mini Rev Med Chem 8:740-754.

[123] Edwards, Mark R. Food, energy and habitat for the 100 Year Starship, NASA, 100 Year Starship Symposium, Orlando, FL, 2011.

[124] Megasun.bch.umontreal.ca/protists/gallery.html algaebase.org/links/ utex.org; ccmp.bigelow.org; http://www.ccap.ac.uk; marine.csiro.au/microalgae; wdcm.nig.ac.jp/hpcc.html).

[125] Graham, Linda and Lee Wilcox. *Algae*. New Jersey: Prentice Hall, 2000: 8.
[126] Hu, Qiang. "Environmental Effects on Cell Composition." *Handbook of Microalgal Culture: Biotechnology and Applied Phycology*. Ed. Amos Richmond. Oxford, England: Blackwell Science, Ltd., 2004: 83-94.
[127] Edwards, Mark R. *Green Solar Gardens*, 12.
[128] Edwards, Mark R. *Smartcultures: Nature's tiny Genius – Algae – Reverses Pollution and Regenerates Degraded Ecosystems*, CreateSpace, 2010, 14.
[129] Perry, Ann. Algae: A Mean, Green Cleaning Machine, *Agricultural Research Magazine*, 58:5, May/June 2010.
[130] http://www.parkseed.com/gardening/PD/9261/
[131] www.cooksgarden.com
[132] Oswald, W.J. and C.G. Golueke, Biological transformation of solar energy, *Advances in. Applied Microbiology*, 2, 1960, 223–262.
[133] Borowitzka, Michael A. Culturing Microalgae in Outdoor Ponds, in Andersen, Robert A., Ed. *Algal culturing techniques*, Phycological Society of America, Elsevier Academic Press, 2005, 205 – 219.
[134] Ugwu CU, Aoyagi H, Uchiyama H. Photobioreactors for mass cultivation of algae. *Bioresource Technology*, 2008, 99(10): 4021-8.
[135] Florida Department of Environmental Protection, www.bioreactor.org/
[136] Edwards, Mark. Unpublished survey research, ASU, 2008.
[137] Behrens, Paul W. Photobioreactors and Fermentors: The Light and Dark Sides of Growing Algae, in Andersen, A., Ed. *Algal culturing techniques*, Phycological Society of America, Elsevier Academic Press, 2005, 189 -205.
[138] Vaidyanathan, Gayathri. Genetic Engineering No Match for Evolution of Weed Resistance, *Scientific American*, April 14, 2010.
[139] Tisdale, S. L. and W. L. Nelson. *Soil Fertility and Fertilizers*. 3rd ed. New York: Macmillan, 1975.
[140] Brady, N. C. *The Nature and Properties of Soils*. New York: Macmillan Publishing Co., 1974.
[141] Gliessman, Stephen R. Agroecology: The Ecology of Sustainable Food Systems, Second Edition, CRC Press; 2 ed., 2006.
[142] Plaster, E. J. *Soil Science and Management*. 3rd ed. Albany: Delmar Publishers, 1996.
[143] Edwards, Mark R. BioWar I, 84.
[144] Prince Charles, Future of Food Conference, Georgetown University May 4, 2011. http://washingtonpostlive.com/conferences/food
[145] DOE, National Algal Technology Roadmap, 2010. http://www1.eere.energy.gov/biomass/pdfs/algal_biofuels_roadmap.pdf

[146] United Nations Environment Programme, UNEP reports. http://www.footprintstandards.org
[147] Brown, Marilyn A., et al. Shrinking The Carbon Footprint of Metropolitan America. Brookings Institution Metropolitan Policy Program, 23 Feb. 2011.
[148] Edwards, Mark R. *Crash! The Demise of fossil Foods and the Rise of Abundance*, Tempe: CreateSpace, 2009, 26.
[149] Hawkesworth, Sophie et. al. Feeding the world healthily: the challenge of measuring the effects of agriculture on health. Phil. Trans. R. Soc. B. September 27, 2010 365:3083-3097;doi:10.1098/rstb.2010.0122
[150] http://www.nrcs.usda.gov/technical/NRI/2007/nri07erosion.html
[151] Avery A. and Avery D. Beef Production and Greenhouse Gas Emissions. *Environmental Health Perspectives*, 2008. 116:A374-A375.
[152] Thelen, Kurt D. What is the Direct Carbon Footprint of Biofuel Relative to Gasoline? Michigan State University, *Crop & Soil Sciences*, https://www.msu.edu/~thelenk3/Acrobat/C%20footprint.pdf
[153] Pimentel, David and Pimentel, Marcia. Food, Energy and Society, 3rd Ed., New York, CRC Press, 2007, 201.
[154] Edwards, Mark R. *Green Algae Strategy*, 144.
[155] Subhadra, Bobban G. and Mark R. Edwards, Coproduct market analysis and water footprint of simulated commercial algal biorefineries, *Applied Energy* 88, 2011, 3515–3523.
[156] Oxford University. http://dx.plos.org/10.1371/journal.pone.0043909
[157] Sydenham E, et al. Omega 3 fatty acid for the prevention of cognitive decline and dementia. Cochrane Database, 2012 Jun 13;6: CD005379.
[158] http://www.worldwater.org
[159] http://sustainablep.asu.edu
[160] Robert Henrikson's blog, Microfarms, *Algae Industry Magazine*.
[161] Henrikson, Robert and Mark Edwards, *Imagine Our Algae Future*, CreateSpace, 2012, 42.
[162] McGovern, George. *Ending Hunger Now*, Fortress Press, 2005, 23.

www.ingramcontent.com/pod-product-compliance
Lightning Source LLC
Chambersburg PA
CBHW061507180526
45171CB00001B/75